The Irrationals

The Irrationals

Julian Havil

PRINCETON UNIVERSITY PRESS
PRINCETON AND OXFORD

Copyright © 2012 by Princeton University Press

Published by Princeton University Press,
41 William Street, Princeton, New Jersey 08540

In the United Kingdom: Princeton University Press,
6 Oxford Street, Woodstock, Oxfordshire OX20 1TW

All Rights Reserved

ISBN: 978-0-691-14342-2 (alk. paper)

Library of Congress Control Number: 2012931844
British Library Cataloguing-in-Publication Data is available

This book has been composed in LucidaBright
Typeset by T&T Productions Ltd, London
Printed on acid-free paper ∞
press.princeton.edu
Printed in the United States of America

3 5 7 9 10 8 6 4

This book is dedicated to our world's most significant tree

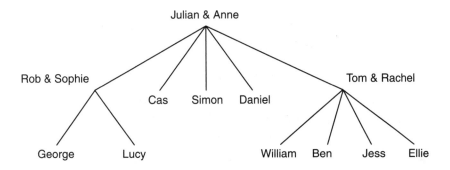

It was difficult enough being a mathematician, this being the frightening subject of which even educated people knew nothing, not even what it was, and of which they might proudly boast ignorance.

Andrew Hodges (of Alan Turing)

A mathematician learns more and more about less and less, until he knows everything about nothing; whereas a philosopher learns less and less about more and more, until he knows nothing about everything.

Anonymous

Contents

Acknowledgments   ix

Introduction   1

CHAPTER ONE
Greek Beginnings   9

CHAPTER TWO
The Route to Germany   52

CHAPTER THREE
Two New Irrationals   92

CHAPTER FOUR
Irrationals, Old and New   109

CHAPTER FIVE
A Very Special Irrational   137

CHAPTER SIX
From the Rational to the Transcendental   154

CHAPTER SEVEN
Transcendentals   182

CHAPTER EIGHT
Continued Fractions Revisited   211

CHAPTER NINE
The Question and Problem of Randomness   225

CHAPTER TEN
One Question, Three Answers   235

CHAPTER ELEVEN
Does Irrationality Matter?   252

APPENDIX A
The Spiral of Theodorus   272

APPENDIX B
Rational Parameterizations of the Circle   278

APPENDIX C
Two Properties of Continued Fractions   281

APPENDIX D
Finding the Tomb of Roger Apéry   286

APPENDIX E
Equivalence Relations   289

APPENDIX F
The Mean Value Theorem   294

Index   295

Acknowledgments

It is a pleasure to recognize the help of my editor, Vickie Kearn, for her unstinting patience and understanding; Jon Wainwright, who typeset the book so expertly and sensitively; Sean Kureshi, Andrew Leigh and George Watkinson for their expertise in computer graphics, ancient Greek and phi from a sound engineer's point of view respectively. And finally a very special mention to Ardavan Afshar and Archie Bott, two former students who read the manuscript so carefully and freely commented on it; by so doing they were of inestimable help.

The Irrationals

Introduction

Trial on air quashed as unsound (10)

1 Down, *Daily Telegraph* crossword 26,488, 1 March 2011

Irrational numbers have been acknowledged for about 2,500 years, yet properly understood for only the past 150 of them. This book is a guided tour of some of the important ideas, people and places associated with this long-term struggle.

The chronology must start around 450 B.C.E. and the geography in Greece, for it was then and there that the foundation stones of pure mathematics were laid, with one of them destined for highly premature collapse. And the first character to be identified must be Pythagoras of Samos, the mystic about whom very little is known with certainty, but in whom pure mathematics may have found its earliest promulgator. It is the constant that sometimes bears his name, $\sqrt{2}$, that is generally (although not universally) accepted as the elemental irrational number and, as such, there is concord that it was this number that dislodged his crucial mathematical–philosophical keystone: positive integers do not rule the universe. Yet those ancient Greeks had not discovered irrational numbers as we would recognize them, much less the symbol $\sqrt{2}$ (which would not appear until 1525); they had demonstrated that the side and diagonal of a square cannot simultaneously be measured by the same unit or, put another way, that the diagonal is *incommensurable* with any unit that measures the side. An early responsibility for us is to reconcile the incommensurable with the irrational.

This story must begin, then, in a predictable way and sometimes it progresses predictably too, but as often it meanders along roads less travelled, roads long since abandoned or concealed in the dense undergrowth of the mathematical monograph. As the pages turn so we unfold detail of some of the myriad results which have shaped the history of irrational numbers, both great and small, famous and obscure, modern and classical – and these last we give in their near original form, costly though that can be. Mathematics can have known no greater aesthete than G. H. Hardy, with one of his most widely used quotations[1]:

> There is no permanent place in the world for ugly mathematics.

Perhaps not, but it is in the nature of things that first proofs are often mirror-shy.[2] They should not be lost, however, and this great opportunity has been taken to garner some of them, massage them a little, and set them beside the approaches of others, whose advantage it has been to use later mathematical ideas.

At journey's end we hope that the reader will have gained an insight into the importance of irrational numbers in the development of pure mathematics,[3] and also the very great challenges sometimes offered up by them; some of these challenges have been met, others intone the siren's call.

What, then, is meant by the term *irrational number*? Surely the answer is obvious:

> It is a number which cannot be expressed as the ratio of two integers.

Or, alternatively:

> It is a number the decimal expansion of which is neither finite nor recurring.

Yet, in both cases, irrationality is defined in terms of what it is not, rather like defining an odd number to be one that is not even. Graver still, these answers are fraught with limitations: for example, how do we use them to define equality between, or arithmetic operations on, two irrational numbers? Although these are familiar, convenient and harmless definitions, they are quite useless in practice. By them, irrational numbers are being defined in terms of one of their characteristic qualities, not as entities in their own right. Who is to say that they exist at all? For novelty, let us adopt a third, less well-known approach:

Since every rational number r can be written

$$r = \frac{(r-1) + (r+1)}{2},$$

---

[1] *A Mathematician's Apology* (Cambridge University Press, 1993).
[2] As indeed was Hardy.
[3] Even if they have no accepted symbol to represent them.

every rational number is equidistant from two other rational numbers (in this case $r - 1$ and $r + 1$); therefore, no rational number is such that it is a different distance from all other rational numbers.

With this observation we define the irrational numbers as:

The set of all real numbers having different distances from all rational numbers.

With its novelty acknowledged, the list of limitations of the definition is as least as long as before. It is an uncomfortable fact that, if we allow ourselves the integers (and we may not), a rigorous and workable definition of the rational numbers is quite straightforward, but the move from them to the irrational numbers is a problem of quite another magnitude, literally as well as figuratively: the set of rational numbers is the same size as the set of integers but the irrational numbers are vastly more numerous. This problem alone simmered for centuries and analysis waited ever more impatiently for its resolution, with the nineteenth-century rigorists posing ever more challenging questions and ever more perplexing contradictions, following Zeno of Elea more than 2,000 years earlier. In the end the resolution was decidedly Germanic, with various German mathematicians providing three near-simultaneous answers, rather like the arrival of belated buses. We discuss them in the penultimate chapter, not in the detail needed to convince the most skeptical, for that would occupy too many pages with tedious checking, but we hope with sufficient conviction for hand-waving to be a positive signal.

For whom, then, is this story intended? At once to the reader who is comfortable with real variable calculus and its associated limits and series, for they might read it as one would read a history book: sequentially from start to finish. But also to those whose mathematical training is less but whose curiosity and enthusiasm are great; they might delve to the familiar and sometimes the new, filling gaps as one might attempt a jigsaw puzzle. In the end, the jigsaw might be incomplete but nonetheless its design should be clear enough for recognition. In as much as we have invested great effort in trying to explain sometimes difficult ideas, we must acknowledge that the reader must invest energy too. Borrowing the words of a former president of Princeton University, James McCosh:

> The book to read is not the one that thinks for you but the one that makes you think.[4]

The informed reader may be disappointed by the omission of some material, for example, the base φ number system, Phinary (which makes essential use of the defining identity of the Golden Ratio), and Farey sequences and Ford Circles, for example. These ideas and others have been omitted by design and undoubtedly there is much more that is missing by accident, with the high ideal of writing comprehensively diluted to one that has sought simply to be representative of a subject which is vast in its age, vast in its breadth and intrinsically difficult. Each chapter of this book could in itself be expanded into another book, with each of these books divided into several volumes.

We apologize for any errors, typographic or otherwise, that have slipped through our mesh and we seek the reader's sympathy with a comment from Eric Baker:

> Proofreading is more effective after publication.

---

[4]He continued: "No book in the world equals the Bible for that." That acknowledged, we regard the sentiment as wider.

The moderation of men gaoled for fiddling pension at last (6,4)
3 Down, *Daily Telegraph* crossword 26,501, 16 March 2011

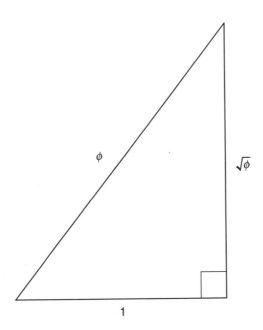

Pythag𝜙ras and the w𝜙rld's m𝜙st irrati𝜙nal number

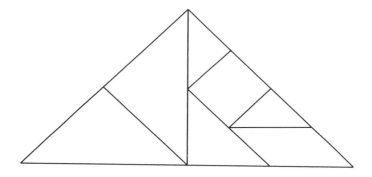

Pythagoras, $\sqrt{2}$ and tangrams

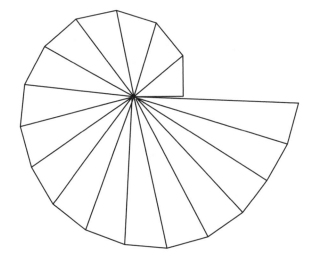

The Spiral of Theodorus

$$\lim_{m \to \infty} \lim_{n \to \infty} \cos^{2n}(m!\pi x) = \begin{cases} 1: & x \text{ is rational} \\ 0: & x \text{ is irrational} \end{cases}$$

Greek Beginnings

It is terrifying to think how much research is needed to determine the truth of even the most unimportant fact.

Stendhal

Sources and Apologia

The birth of irrational numbers took place in the cradle of European mathematics: the Greece of several centuries B.C.E. For this conviction and for much more we must place reliance on a few fragments of contemporary papyrus, some complete but much later manuscripts, and the scholarship of many specialists who, even between themselves, sometimes disagree in fundamental ways. Of great importance is the following passage:

> Thales,[1] who had travelled to Egypt, was the first to introduce this science [geometry] into Greece. He made many discoveries himself and taught the principles for many others to his successors, attacking some problems in a general way and others more empirically. Next after him Mamercus, brother of the poet Stesichorus, is remembered as having applied himself to the study of geometry; and Hippias of Elis records that he acquired a reputation in it. Following upon these men, Pythagoras transformed mathematical philosophy into a scheme of liberal education, surveying its principles from the highest downwards and investigating its theorems in an immaterial [abstract] and intellectual manner. He it was who discovered the doctrine of proportionals and the structure of the cosmic figures.[2]

---

[1] Thay-leez, but originally T-hay-leez.

[2] From the translation by Glenn R. Morrow: *Proclus, A Commentary on the First Book of Euclid's Elements* (Princeton University Press, 1992).

So begins the *Eudemian Summary*, which forms part of the second of two prologues to *A Commentary on the First Book of Euclid's Elements*, written by Proclus (411–485 C.E.), or to give him his full accepted name, Proclus Diadochus (Proclus the Successor), for reasons we shall soon discover. Here Proclus, the last great ancient Greek philosopher, mentions Thales of Miletus (624–546 B.C.E.), who was possibly the first, as he might have been the first pure mathematician. Proclus also mentions Pythagoras of Samos (580–520 B.C.E.), who might have been the second, and with whom the story of irrational numbers, according to what evidence is available, should begin. Our reliance on Proclus[3] is not novel and we will call on the distinguished British polymath Ivor Bulmer-Thomas to provide a proper historical perspective:[4]

> It [the *Eudemian Summary*] is, along with the Collection of Pappus and the commentaries of Eutocius on Archimedes, one of the three most precious sources for the early history of Greek mathematics. In his closing words Proclus expressed the hope that he would be able to go through the remaining books [of Euclid's *Elements*] in the same fashion; there is no evidence that he ever did so, but as Book 1 contains definitions, postulates, and axioms underlying all the remainder we have the most important things that he would have wished to say.

We have, then, a few pages of observations written a thousand years after the events they chronicle as a principal source of reliable information about these ancient times, and these take the form of a commentary on part of another paramount source; the most influential, most studied, most copied,[5] and most widely read mathematical work ever to be written: Euclid's *Elements*. Ironically, the paucity of surviving written material is in no small part due to this iconic work, without which our knowledge of ancient Greek mathematics would be so significantly impoverished. The Greeks' medium of record was papyrus, made from

---

[3]"Proclus is to Euclid as Boswell is to Johnson", according to Howard Eves.

[4]Ivor Bulmer-Thomas, 1972, Proclus on Euclid I, *The Classical Review* (New Series) 22:345–47.

[5]For novelty, the reader may wish to peruse Oliver Byrne's remarkable edition, which is now available in reprint: Werner Oechslin, 2010, *Byrne, Six Books of Euclid*, ed. Petra Lamers-Schutze (Harper Paperback).

a grass-like plant originating in the Nile delta and subject to natural and often rapid decomposition, particularly in the comparatively damp Greek climate. In that climate, it was simply not a safe long-term medium of record: it rotted. Those works which were deemed worthy of the considerable expense of being preserved were copied by scribes, perhaps as faithful replicas or perhaps with changes that were thought to be appropriate; the remainder were simply left to decompose. From the same reference, again we hear from Ivor Bulmer-Thomas:

> Euclid's *Elements* was so immediately successful that it drove all its predecessors out of the field.

It is a simple fact that the prominence of *The Elements* reduced to irrelevance much that preceded it, condemning the works to obscurity and then oblivion; David Hilbert's remark, made in the nineteenth century, really sums up the situation quite nicely:

> One can measure the importance of a scientific work by the number of earlier publications rendered superfluous by it.

We shall have much need of *The Elements* later in this chapter but the reader should be clear that we must again rely on secondary sources, since no extant version of it exists. In fact, the earliest surviving copies of the work, held at the Vatican and the Bodleian Library in Oxford, date from the ninth century C.E.; a thousand years after Euclid. That said, some much earlier fragments have been found on potsherds discovered in Egypt dating from around 225 B.C.E. and pieces of papyrus dated from 100 B.C.E.: the former containing notes on two Propositions from Book XIII, and the latter having inscribed parts of Book II.

So, with the Greek's medium of record fatally inadequate, *The Elements* relegating untold numbers of earlier works to insignificance, and the effect of providence that accompanies the passing of so many years, we must accept the consequent historical difficulties: and these do not end here. Really, they have their beginning with the Greek custom, until about 450 B.C.E., of transmitting knowledge orally, continue with the fondness on the part of later commentators to exaggerate the contributions of great men and culminate with the staged destruction of the academic riches of Alexandria: the Romans (seemingly in 48 B.C.E.) razed the great Library of Alexandria with its estimated 500,000 manuscripts, the

Christians (in 392 C.E.) pillaged Alexandria's Temple of Serapis with its possible 300,000 manuscripts, and finally the Muslims burnt thousands more of its books (in about 640 C.E.). Add to these the Pythagorean custom of attributing all results to their founder, who appears never to have written anything down, and their (near) strict adherence to their canon of omerta, and we have the ingredients of the mathematical historian's nightmare; judging veracity and objectivity of the scant available evidence is a responsibility properly undertaken only by a few specialists, upon whom our discussion must rely.

Specifically, as to our knowledge of Pythagoras, the contemporary classicist Professor Carl Huffman provides a dampening perspective[6]:

> ...any chronology constructed for Pythagoras' life is a fabric of the loosest possible weave.

He may have been a pupil of Thales and to Proclus can be added significant further material; for example, Plato (428–347 B.C.E.) mentions him as a great teacher in Book X of *The Republic*, which is dated somewhere around 380 B.C.E. And there are three biographies too: that of Diogenes Laertius (200–250 C.E.), who wrote as part of a ten-volume work on the lives of Greek philosophers, the volume *Life of Pythagoras*; the other two are those of Iamblichus of Chalcis (ca. 245–325 C.E.), *On the Pythagorean Way of Life*, and of his teacher, Porphyry (234–305 C.E.) with *Life of Pythagoras*. All were written about eight hundred years after Pythagoras' time, but they are at least extant. The definitive modern treatise must surely be that by the German classical scholar Walter Burkert,[7] to which we refer the interested and committed reader; we shall be content with the following thumbnail impression of Pythagoreanism, which will suit our own modest needs.

All things are number: such was the Pythagorean dictum central to their philosophy. To them, number meant the discrete positive integers, with 1 the unit by which all other numbers were

---

[6]Huffman, Carl, 1993, *Philolaus of Croton Pythagorean and Presocratic: A Commentary on the Fragments and Testimonia with Interpretive Essays* (Cambridge University Press), pp. 1–16.

[7]*Lore and Science in Ancient Pythagoreanism*, translated by Edwin Minar (Harvard University Press, 1972).

measured. This meant that all pairs of numbers were each multiples of the unit; that is, all pairs of numbers were *commensurable* by it. Contrastingly, lengths, areas, volumes, masses, etc., were continuous quantities, the *magnitudes* which served the ancient Greeks in place of real numbers. Ratios of discrete were conceptually secure and those of magnitudes could be envisaged too, provided that the two values concerned were of the same type. Further, the modern statement

$$A:B = C:D$$

was meaningful, where on one side of the equality there are magnitudes of one type and on the other magnitudes of another type. This could mean lengths on one side and areas on the other, and we will see part of the utility of this a little later. Additionally, their study of musical scales revealed that philosophical coincided with musical harmony with cordant sounds found to be measured by integer ratios of lengths of strings with, for example, the octave corresponding to a ratio of length of 2 to 1 and a perfect fourth to 3 to 2, etc. This was evidence to them that the continuous could be measured by the discrete. We will allow Aristotle[8] to summarize the situation:

> Contemporaneously with these philosophers and before them, the so-called Pythagoreans, who were the first to take up mathematics, not only advanced this study, but also having been brought up in it they thought its principles were the principles of all things. Since of these principles numbers are by nature the first, and in numbers they seemed to see many resemblances to the things that exist and come into being – more than in fire and earth and water (such and such a modification of numbers being justice, another being soul and reason, another being opportunity – and similarly almost all other things being numerically expressible); since, again, they saw that the modifications and the ratios of the musical scales were expressible in numbers; since, then, all other things seemed in their whole nature to be modelled on numbers, and numbers seemed to be the first things in the whole of nature, they supposed the

---

[8] Aristotle, Metaphysics, Book 1(1), translated by W. D. Ross, available at http://ebooks.adelaide.edu.au/a/aristotle/metaphysics/complete.html (accessed 19 September 2011).

elements of numbers to be the elements of all things, and the
whole heaven to be a musical scale and a number. And all the
properties of numbers and scales which they could show to
agree with the attributes and parts and the whole arrangement
of the heavens, they collected and fitted into their scheme; and
if there was a gap anywhere, they readily made additions so
as to make their whole theory coherent. E.g. as the number
10 is thought to be perfect and to comprise the whole nature
of numbers, they say that the bodies which move through the
heavens are ten, but as the visible bodies are only nine, to meet
this they invent a tenth – the 'counter-earth'.

With the Pythagoreans' dogmatism, the stage was set for the crisis
in Greek mathematics that was to unfold, as 'a veritable logical
scandal',[9] possibly the very first of the long sequence of them
that continues to this day.

So, there are available sources and there are scholars who have
mined them of their dependable evidence regarding these remote
times. By now, however, we hope that the reader will have a
proper appreciation of the intrinsic historical complications and
accept that given dates are sometimes approximate and statements made only with the authority borne of a compromise of
accepted wisdom.

As a final emphasis we can gain some idea of the chain con-
necting Thales and Pythagoras to Proclus by relying on the schol-
arly labour of the early twentieth-century Dutch mathematical
researcher J. G. van Pesch,[10] which includes a detailed study of
the works which he deemed were accessible to and directly used
by Proclus, whether or not the dependence was explicitly stated.
Figure 1.1 shows the resulting timeline of individuals and consists
of a mixture of familiar and not-so-familiar names, together with
approximate dates. Most particularly, the name of Eudemus of
Rhodes (350–290 B.C.E.) appears, an historian who is attributed
with writing a long-lost history of Greek geometry covering the
period prior to 335 B.C.E.; it is, in particular, this formative work
that van Pesch (and others) are confident that Proclus had at his
disposal and summarized: hence *Eudemian Summary*.

[9]Paul Tannery, 1887, *La Géométrie Grecque* (Paris), pp. 141–61.

[10]*De Procli fontibus*, Dissertatio ad historiam mathemsecs Graecae pertinens
(Lugduni-Batavorum, Apud L. Van Nifterik, 1900).

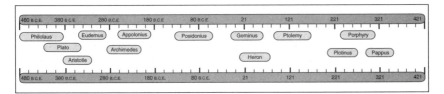

Figure 1.1.

Yet, two names on van Pesch's list are missing from the time-line, since the dates of these individuals are simply not known. The first is Carpus of Antioch, or 'Carpus the Engineer', to whom Proclus attributed the definition *that an angle is a quantity, specifically, the distance between the containing lines or planes*; he is also the person accredited by Iamblichus of Chalcis as being among the Pythagoreans who solved one of the three great problems of antiquity: that of the impossibility of squaring the circle. He appears to have lived at some time between 200 B.C.E. and 200 C.E.

The second name is that of Syrianus of Alexandria, himself a commentator on Plato and Aristotle, and through him we can learn a little about Proclus himself. The revival of Plato's academy under the leadership of Plutarch (46–120 C.E.) brought important scholars to Athens, among whom was the philosopher Syrianus and a young man, about 20 years old and of immense promise, by the name of Proclus. Such was his promise that for a short time before his death the aged Plutarch agreed, exceptionally, to tutor Proclus. When Syrianus replaced Plutarch as leader he also replaced him as tutor to Proclus. In his turn, Proclus succeeded Syrianus as head of the academy; it was this event that brought about the addition of Diadochus (Successor) to his name. The relationship between Proclus and Syrianus was to become intellectually and emotionally immensely close, so much so that Proclus left instructions that, on his death, he be placed in the tomb already occupied by Syrianus. The tomb was located on the slopes of the Lycabettus Hill, overlooking Athens, a limestone peak of some 1000 feet and, for very good reasons, a modern tourist attraction. The level of affection is easily judged by Proclus' decree that the following epitaph be inscribed on their joint resting place:

> Proclus was I, of Lycian race, whom Syrianus
> Beside me here nurtured as a successor in his doctrine.
> This single tomb has accepted the bodies of us both;
> May a single place receive our two souls.

As with so much else, the tomb has long ago disappeared.

Inheritance and Legacy

With the authority of Proclus, Thales, the first of the Seven Sages of Greek tradition, brought geometry from Egypt to Greece. In particular, as the Commentary develops, Proclus attributes to him the following four geometric results:

1. A circle is bisected by any diameter.
2. The base angles of an isosceles triangle are equal.
3. The angles between two intersecting straight lines are equal.
4. Two triangles are congruent if they have two angles and one side equal.

These seem modest achievements. Yet their simplicity belies their significance, as they exhibit the germ of the deductive procedures of Greek philosophy being brought to bear on mathematical processes: the yet more ancient Egyptian and Babylonian civilizations had no thought of axiomatics, abstraction or generalization, with mathematical results having the form of mysterious individual recipes. This is not to say that some of the knowledge could not be called 'advanced', it is simply that the deductive method that we consider as an essential aspect of pure mathematics was entirely absent. Paradoxically, evidence abounds regarding these older civilizations; the ancient Egyptians used papyrus too, but their climate was more papyrus-friendly than Greece, and the Babylonians wrote in cuneiform on wet clay tablets, thousands of which have survived.

To gain a perspective of the magnitude of the step that had been taken by Thales we will trouble to annotate two contemporary examples, one from each of these civilizations.[11] From the Moscow papyrus, which dates from around 1850 B.C.E., we have an Egyptian problem:

Method of calculating a △
If you are told △ of 6 as height, of 4 as lower side,
   and of 2 as upper side.
You shall square these 4. 16 shall result.
You shall double 4. 8 shall result.
You shall square these 2. 4 shall result.
You shall add the 16 and the 8 and the 4. 28 shall result.
You shall calculate $\bar{3}$ of 6. 2 shall result.

---

[11] Victor J. Katz (editor), *The Mathematics of Egypt, Mesopotamia, China, India and Islam* (Princeton University Press, 2007).

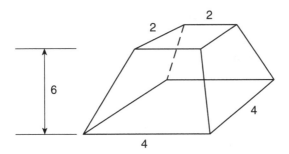

Figure 1.2.

You shall calculate 28 times 2. 56 shall result.
Look, belonging to it is 56.
What has been found by you is correct.

Here the $\overline{3}$ is our $\frac{1}{3}$, which means that the calculation is $\frac{1}{3} \times 6 \times$ $(4^2 + 4 \times 2 + 2^2) = 56$, a special case of the general formula for the volume of the frustram of a square pyramid, $V = \frac{1}{3}h(a^2 + ab + b^2)$, with this special case shown in figure 1.2.

And from a Babylonian tablet from about 2000–1600 B.C.[12]:

A circle was 1 00.
I descended 2 rods.
What was the dividing line (that I reached)?
You: ≪you≫ ≪Square≫ <double> 2.
You will see 4.
Take away ≪you will see≫ 4 from 20, the dividing line.
You will see 16.
Square 20, the dividing line.
You [will see] 6 40.
Square 16.
You will see 4 16.
Take away 4 16 from 6 40.
You will see 2 24.
What squares 2 24?
12 squares it, the dividing line.
That is the procedure.

The Babylonians used base 60. With that in mind, the instructions refer to figure 1.3, a circle of circumference 1 00 in base 60

---

[12]Here the translator has used the conventions: [restored missing words], <restored accidental omissions>, ≪accidental inclusions≫ and (editorial glosses)).

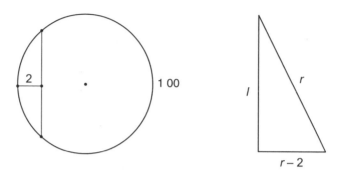

Figure 1.3.

and so 60 units in base 10 and describe a procedure for calculating the length of the chord ('the dividing line that I reached')
2 units in from the circle itself. The mysterious 20 emerges from
the fact that, if the circle has radius r, $2\pi r = 60$ and, with π
taken as 3, $2r = 20$ ($\frac{25}{8}$ was another approximation for π).

In modern notation, with a chord of length $2l$ we have

$$(2l)^2 = (2r)^2 - (2r - 2^2)^2,$$

which simplifies to

$$l^2 = r^2 - (r - 2)^2,$$

and this is no more than an application of Pythagoras's Theorem,
at least a thousand years before Pythagoras.

These are typical of their kind, with the underlying principles entirely hidden and no hint of general results being appreciated. The great seventeenth-century German philosopher
Immanuel Kant encapsulated the transition from Egyptian–Babylonian instruction sets to Greek mathematics in the following
manner[13]:

> In the earliest times to which the history of human rea-
> son extends, mathematics, among that wonderful people, the
> Greeks, had already entered upon the sure path of science... A
> new light flashed upon the mind of the first man (be he Thales
> or some other) who demonstrated the properties of the isosce-
> les triangle. The true method, so he found, was not to inspect

---

[13]Preface to the second edition of the *Critique of Pure Reason* (1929, Norman
Kemp Smith translation).

what he discerned either in the figure, or in the bare concept of it, and from this, as it were, to read off its properties; but to bring out what was necessarily implied in the concepts that he had himself formed a priori, and had put into the figure in the construction by which he presented it to himself.

With this, the direction of mathematical progress was determined as that of a science: from 'reasonable' assumptions deduce logical conclusions. The infamous *parallel postulate* of Euclid serves as the second example of such an assumption concealing within it formative complications: the assumption of universal commensurability is the first.

The Unmentionable Incommensurable

Thales may have started mathematics on what Kant described as the "royal road of science" but tradition has it that it was Pythagoras who guided it along that path, not very far along which lay the concept of *incommensurability*, the Greeks' perception of what we now call *irrationality*.

Recall that, to the Pythagoreans, all things were measured by number; that is, by positive integers, or the ratio of two of them. Recall further that it was their conviction that the word *all* embraced continuous magnitudes such as length, area and volume, which meant that not only were the cordant notes produced by plucking strings so constrained but also the strings which produced those notes: that is, any two lengths were commensurable with each other, not this time with a single unit measuring them but the much more mysterious unspecified and variable unit. That is, given two lines of different lengths (two different magnitudes), for the Pythagoreans there must exist a third line which subdivides both of them perfectly, the length of which is their common unit. (Of course, any subdivision of that line would also be a unit.) In modern notation, if one line is of length l_1 and another of length l_2 and the common unit is u there must exist integers n_1 and n_2 such that $l_1 = n_1 u$ and $l_2 = n_2 u$. The consequence of this is that the ratio of any two magnitudes is the ratio of two integers, $l_1 : l_2 = n_1 u : n_2 u = n_1 : n_2$. And a great deal more than philosophical expediency depends on this outcome, as we shall see.

Unfortunately, it does not appear to have taken long for the logical fault-line associated with this approach to magnitudes to be exposed, but how it was uncovered and by whom are matters of academic debate.

One possibility is that the Pythagorean interest in the concept of the *mean* of two numbers may have directed their attention to the nature of the geometric mean of 1 and 2; 1 the monad which generates all numbers and 2 the dyad and first feminine symbol. The general approach seems to have been to consider two numbers a and c and to define a third number b, which is a *mean* of them, with $a \leqslant b \leqslant c$. The method was to note that each of $b-a$, $c-b$ and $c-a$ is $\geqslant 0$ and that the ratio of any pair of these can be compared with that of any pair of the original numbers, with the three fundamental examples in modern notation:

$$\frac{b-a}{c-b} = \frac{a}{a} = \frac{b}{b} = \frac{c}{c} \quad \text{to yield } b = \frac{a+c}{2}, \text{ the arithmetic mean;}$$

$$\frac{c-b}{b-a} = \frac{b}{a} = \frac{c}{b} \quad \text{to yield } b = \sqrt{ac}, \text{ the geometric mean;}$$

$$\frac{c-b}{b-a} = \frac{c}{a} \quad \text{to yield } b = \frac{2}{1/a + 1/c}, \text{ the harmonic mean.}$$

With the geometric formula giving

$$\frac{b}{1} = \frac{2}{b}$$

we have the appearance of $\sqrt{2}$.

This said, tradition has it that the author of the destruction of the Pythagorean ideal was none other than an acolyte: Hippasus of Metapontum, a man who has a decidedly negative Pythagorean press. Not only is he accused of destroying the concept of commensurability, he is meant to have spoken of the horror outside the secretive Pythagorean community – and he is meant to have done the same with his discovery that a dodecahedron can be inscribed within a sphere. With the authority of Iamblichus of Chalcis, whom we mentioned earlier and who was to have a considerable influence on Proclus[14]:

---

[14] John Dillon (editor) and Jackson Hershbell (translator), 1991, *Iamblichus: On the Pythagorean Way of Life. Text, Translation, and Notes* (Atlanta: Scholars Press).

They say that the man who first divulged the nature of commensurability and incommensurability to men who were not worthy of being made part of this knowledge, became so much hated by the other Pythagoreans, that not only they cast him out of the community; they built a shrine for him as if he were dead, he who had once been their friend. Others add that even the gods became angry with him who had divulged Pythagoras' doctrine; that he who showed how the dodecahedron can be inscribed within a sphere died at sea like an evil man. Others still say that the same misfortune happened on him who spoke to others of irrational numbers and incommensurability.

The Pythagoreans may, according to Aristotle, have been able to invent a 'counter-earth' to fit in with their cosmological views but incommensurability was not so easily disposed of. The ratio of two magnitudes was now not necessarily defined, and this meant that the fundamental geometric tool of similarity was denied them – and all results depending on its use, as we shall see.

Mention of the phenomenon is made by Plato in several of his Dialogues, and these reveal an interesting linguistic development: in the *Hippias Major* (if indeed he was its author) and *The Republic* incommensurables are referred to as $\alpha\rho\rho\varepsilon\tau o\varsigma$, or *unmentionable*, whereas in the later Dialogues, *Theaetetus* and *Laws* the change is to $\mu\chi\varepsilon\iota o\sigma\mu\mu\varepsilon\tau\rho o\iota$, or *incommensurable*. We shall soon see how this mutated to *surd*.

Whether it was Hippasus or another who discovered incommensurability and whether or not this was achieved through the study of means is not known; the method of proof remains yet another mystery but it is popular tradition that he applied his master's theorem to a square of side 1 unit, somehow showing that its diagonal of length $\sqrt{2}$ was incommensurable with the unit side. Such an argument appears in Book X, Appendix 27, of *The Elements*, as shown below[15]:

Let ABCD be a square and AC its diameter. I say that AC is incommensurable with AB in length. For let us assume that it is commensurable. I say that it will follow that the same number is at the same time even and odd. It is clear that the square on AC is double the square on AB. Since then (according to

[15]Kurt Von Fritz, 1945, The discovery of incommensurability by Hippasus of Metapontum, *The Annals of Mathematics* (Second Series) 46(2):242–64.

our assumption) AC is commensurable with AB, AC will be to AB in the ratio of an integer to an integer. Let them have the ratio DE:DF and let DE and DF be the smallest numbers which are in this proportion to one another. DE cannot then be the unit. For if DE was the unit and is to DF in the same proportion as AC to AB, AC being greater than AB, DE, the unit, will be greater than the integer DF, which is impossible. Hence DE is not the unit, but an integer (greater than the unit). Now since AC:AB = DE:DF, it follows that also $AC^2 : AB^2 = DE^2 : DF^2$. But $AC^2 = 2AB^2$ and hence $DE^2 = 2DF^2$. Hence DE^2 is an even number and therefore DE must also be an even number. For, if it was an odd number, its square would also be an odd number. For, if any number of odd numbers are added to one another so that the number of numbers added is an odd number the result is also an odd number. Hence DE will be an even number. Let then DE be divided into two equal numbers at the point G. Since DE and DF are the smallest numbers which are in the same proportion they will be prime to one another. Therefore, since DE is an even number, DF will be an odd number. For, if it was an even number, the number 2 would measure both DE and DF, although they are prime to one another, which is impossible. Hence DF is not even but odd. Now since DE = 2EG it follows that $DE^2 = 4EG^2$. But $DE^2 = 2DF^2$ and hence $DF^2 = 2EG^2$. Therefore DF^2 must be an even number, and in consequence DF is also an even number. But it has also been demonstrated that DF must be an odd number, which is impossible. It follows, therefore, that AC cannot be commensurable with AB, which was to be demonstrated.

A simple sketch and a little patience reveal the argument which, in modern algebraic form, is greatly familiar.

Suppose that $\sqrt{2} = a/b$, where a and b are integers in their lowest terms. Then $a^2 = 2b^2$ and so a^2 is even and therefore a is even. Write $a = 2k$, then $a^2 = 4k^2$ and so $2b^2 = 4k^2$ and therefore $b^2 = 2k^2$, which makes b even and we have a contradiction to the assumption that the two numbers are in lowest terms.[16]

In its geometric form the inherent beauty of the argument is concealed but it stands, with the infinitude of the primes, as one

---

[16]Notice, though, that the $a^2 = 2b^2$ stage is sufficient to yield the contradiction if the equation is written in binary: in binary, the square of any positive integer must begin on the right with an even number of 0s (including none of them) and that multiplying by 2 introduces one more.

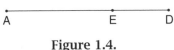

Figure 1.4.

of the two most elegant and elementary proofs of mathematics, each chosen for those reasons by G. H. Hardy for inclusion in his *Mathematician's Apology*.

Yet, there is a sustainable view that this Euclidean neatness is an example of *The Elements* presenting a later rather than an original approach and we will now follow an alternative path which leads to the annihilation of Pythagorean commensurability.[17]

Signs of Danger

It can be of little surprise that, over the ages, commentators have remarked on the mystical significance the Pythagoreans attached to some shapes. For example, if we look to the original Oxford English Dictionary's 1908 entry for *Pythagorean* we see its famous editor, James Murray, allowing the capital Greek letter upsilon (Y) to be their representation of the two divergent paths of virtue and vice. The equally prestigious American initiative, the *Century Dictionary*, has the 1906 entry *Hexagram* attaching the regular hexagon and its associated hexagram to Pythagorean mysticism. Modern versions of these publications and others like them are consistent with their illustrious predecessors. The great classical scholar Sir Thomas Heath, to whom we have already referred, cites Lucian of Samosata (125–180 C.E.) and the scholist to the far earlier play of Aristophanes, *The Clouds*, as authority that the pentagon and associated pentagram were symbols of Pythagorean recognition.

Square, Pentagram, Hexagram – no matter: they each conceal commensurability's doom without the need of Pythagoras's theorem; their own sacred symbols were harbingers of their philosophical destruction.

To consider the matter, first we need to look at an immediate consequence of the definition of commensurability, one which is widely applied in *The Elements*:

> Referring to figure 1.4, if AD and AE are commensurable, then it must be that AD and AD – AE are commensurable.

---

[17]See, for example, Kurt Von Fritz, ibid.

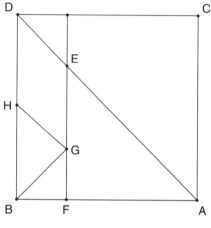

Figure 1.5.

The result is obvious and its proof trivial but, undeterred by either fact, let us suppose that AD and AE are commensurable with a unit u and that AD $= Nu$ and AE $= nu$, then AD $-$ AE $= Nu - nu = (N - n)u$, which indeed means that AD and AD $-$ AE are commensurable by the unit u.

Now consider the square as shown in figure 1.5, with the line AED located as a diagonal, where the point E is defined in the first part of the following construction.

Select the point E on the diagonal AD so that AE = AB and drop the perpendicular EF. Now consider the segment ED and move a copy of it vertically downwards as the segment GH, thereby creating a parallelogram DEGH, and consider what happens to the size of $\angle$BGH. Initially, when GH was the same as ED with $\angle$BGH $= \angle$BED $> 90°$ and finally, when G becomes F, $\angle$BGH $= \angle$BFH $< 90°$. Since the process is continuous, there must be a position of GH so that $\angle$BGH $= 90°$; let that be the position shown in the figure and consider the right-angled isosceles triangle BGH. If AD is commensurable with AB, then it is commensurable with AE and so, using the above observation, AD is commensurable with AD $-$ AE $=$ ED $=$ GH. Now let GH be the diagonal of the next square and the process can be repeated indefinitely with ever smaller nested triangles to ensure that no unit of commensurability can exist. It must be that AD is incommensurable with AB, never mind using Pythagoras's theorem to find the length.

Figure 1.6.

So the simple square, the first of the regular shapes gifted with Pythagorean mystical significance, conceals the means of commensurability's destruction. Now let us look at a second, the regular pentagon; a figure redolent with hidden meaning which gives rise to that Pythagorean symbol of recognition, most commonly known as the pentagram.

The pentagram, pentalpha, pentangle or star pentagon is one of the most potent, powerful and persistent symbols in the history of humankind. It has been throughout the ages a symbol used by pagans, ancient Israelites, Christians, magicians, Wiccans and many more cults. Sir Thomas Mallory in *Le Morte d'Arthur* had Sir Gawain adopt it as his personal symbol to be placed it on his shield and Dan Brown in *The Da Vinci Code* had the dying Louvre museum curator Jacque Saunière draw one on his abdomen in his own blood as a clue to identify his murderer. It can be realized with or without its defining regular pentagon or circumscribing circle, as we see in figure 1.6. For the mathematician it is the simplest example of a *star polygon*: each of five regularly spaced points having been connected by a straight line to another, with the general case having the connection made between every mth point of the n points, with the resulting figure commonly given the symbol $\{n/m\}$. For reasons of common sense and aesthetics it is usually assumed that m and n are relatively prime, $n > 2m$ and that the points are equidistant from the figure's obvious centre; with this notation, the pentagram is given the symbol $\{\frac{5}{2}\}$. Other examples, $\{\frac{8}{2}\}$ $\{\frac{7}{3}\}$ and $\{\frac{6}{2}\}$, are shown in figure 1.7, the last of which will attract our attention in a few lines.

If we consider the pentagram with its defining regular pentagon, the eye is led to its many congruent or similar isosceles triangles and to the nested regular pentagon, inverted and suggestive of an infinitely recursive extension of the figure, the next stage of

Figure 1.7.

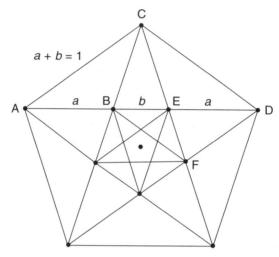

Figure 1.8.

which is shown in figure 1.8. Once again we can force a contradiction in much the same manner as with the square. This time locate the line AED as a diagonal of the pentagon and suppose that the diagonal AD and the side AC of the large pentagon are commensurable. We have AD − AC = AD − AE = ED = AB and AB = BC = BF, with BF a diagonal of the inner pentagon. Since AD and AD − AE = AB = BF are commensurable, AD, the diagonal of the original pentagon must be commensurable with BF, the diagonal of the smaller, nested pentagon. Once again, since the process can be continued indefinitely, whatever the supposed unit that measures both AD and AE it will eventually be bigger than the length of the diagonal of a nested pentagon; inescapably, there can be no such common measure.

Finally (and it must be finally, as we will soon mention), that $\{\frac{6}{2}\}$ star polygon, the regular hexagon, can be used to generate

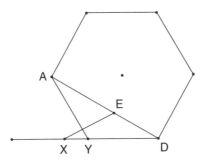

Figure 1.9.

the same infinite process. Its associated double overlapping equilateral triangles, the hexagram, is again one of the oldest and most widespread spiritual symbols, whether it be called the Seal of Solomon, the Star of David, the Shatkona or, from the Hebrew, *the workings of the Seven Planets under the presidency of the Sephiroth and of Ararita, the divine name of Seven letters*; and once again Dan Brown made use of it in that same novel. For the first time the diagonals are not all of the same length, with the longest one evidently commensurable with the side, being twice its length. But what about the shorter diagonals? We now locate the line AED as such a diagonal, as shown in figure 1.9, where E is defined so that AE = AY. Now define point X on DY produced by the condition that ∠AEX = 60°, then XED forms two adjacent sides of a smaller regular hexagon, with shorter diagonal DX. If AD is commensurable with AY then AD is commensurable with AE and so, using the initial result once more, with ED, the side of the new regular hexagon; the infinite process is again started, leading to the demolition of the assumption of commensurability.

In modern terms, the diagonal of the unit square, $\sqrt{2}$, is irrational, as are the diagonals of the regular pentagon and shorter diagonal of the regular hexagon. We can satisfy ourselves which irrationals these last two are in the following manner.

Again, using the notation of figure 1.8 and the symmetry of the pentagram we know that, with a side of 1 unit, AE = AC = $a + b$ = 1, and also that the two triangles ABC and ACD similar. This last fact means that

$$\frac{AC}{AD} = \frac{AB}{AC} \quad \text{and therefore} \quad \frac{1}{a+1} = \frac{a}{1},$$

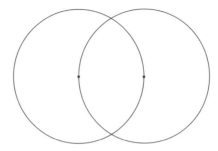

Figure 1.10.

which becomes $a^2 + a - 1 = 0$, and this makes $a = \frac{-1+\sqrt{5}}{2}$ and the length of the diagonal

$$2a + b = 2a + (1 - a) = 1 + a = 1 + \tfrac{1}{2}(-1 + \sqrt{5}) = \tfrac{1}{2}(1 + \sqrt{5}) = \varphi,$$

which is the Golden Ratio, also exposed as the Greek mean of 1 and 2 (as mentioned on page 20), defined by

$$\frac{c - b}{b - a} = \frac{a}{b}.$$

It is a further irony that this 'division into mean and extreme ratio', so revered in Greek mathematics and seemingly so much part of their architecture, should be central to the ruin of its earliest mathematical precept.

Finally, it is a trivial matter to see that the shorter diagonals of the regular hexagon of side 1 unit are of length $\sqrt{3}$.

Quite remarkably, the argument fails thereafter; that is, trying to force the infinite regress for regular polygons of sides greater than 6 is doomed to failure, as shown in a beautiful but lengthy argument by E. J. Barbeau.[18]

So, these important Pythagorean symbols have embodied within them all that is needed to destroy their cherished ideal of commensurability and as if that were not enough so does another of a different type: the Vesica Pisces or *Vessel of the Fish*, which, to the Pythagoreans, was a symbolic womb giving life to the entire universe. The figure is formed by allowing two circles of equal radii to overlap and pass through the centre of the other, as in

---

[18]E. J. Barbeau, 1983, Incommensurability Proofs: A Pattern That Peters Out, *Mathematics Magazine* 56(2):82–90.

figure 1.10, and if the circles have radius 1 then lines of length $\sqrt{2}$, $\sqrt{3}$ (and $\sqrt{5}$) are easily found within the figure, as is φ if we allow the lengths of two of them to be added.

Of course, we cannot be certain whether it was Hippasus or not and whether it was the diagonal of a square or of a pentagram or of a hexagram or whatever that destroyed Pythagorean commensurability, but whoever the architect and whatever his methods, this was to be a revelation that placed a boulder in the royal mathematical path which was not to be circumvented for a century.

Signs of Progress

Plato (429–347 B.C.E.): student and acolyte of Socrates, teacher of Aristotle, author of the Dialogues and founder of the Academy and who, according to the *Stanford Encyclopedia of Philosophy*,

> is, by any reckoning, one of the most dazzling writers in the Western literary tradition and one of the most penetrating, wide-ranging, and influential authors in the history of philosophy.

We have already mentioned that he alluded to incommensurability in several of his Dialogues, the dates of which are matters of dispute, although it seems plain that they began after the suicide of Socrates in 399 B.C.E. and, since Plato died in 347 B.C.E., we have a clear window of time through which to peer. What is not in dispute is the importance of Socrates to them and indeed they form the principal source (along with Xenophon's *Memorabilia of Socrates*) of the great philosopher's achievements; as with Pythagoras, there is a complete absence of anything written by him. Although the exact dates of the Dialogues are questioned, there is reasonable concord concerning the order of them and we will mine three of them from, what might be termed, the Early, Middle and Late periods, for news of progress with irrational numbers.

From the Early *Hippias Major* we learn something of the progress made in the arithmetical properties of rationals and irrationals, with Socrates asking of Hippias[19]:

---

[19]Benjamin Jowett's translation.

To which group, then, Hippias, does the beautiful seem to you
to belong? To the group of those that you mentioned? If I am
strong and you also, are we both collectively strong, and if I
am just and you also, are we both collectively just, and if both
collectively, then each individually so, too, if I am beautiful
and you also, are we both collectively beautiful, and if both
collectively, then each individually? Or is there nothing to pre-
vent this, as in the case that when given things are both collectively even, they may perhaps individually be odd, or perhaps
even, and again, when things are individually irrational quantities they may perhaps both collectively be rational, or perhaps
irrational, and countless other cases which, you know, I said
appeared before my mind?

And then the later Early Transitional Dialogue, *Meno*, has Socrates
asking Meno's boy slave how to find the side of a square the area
of which is double that of a given square. The resulting exchanges
between Socrates and the boy are a model of the Socratic approach
to teaching, with the boy being led from his first incorrect suggestion that the new square should have side double that of the original to the eventual correct answer of the side being the diagonal
of the first square. With this dialogue, what we would call $\sqrt{2}$ is
acknowledged.

But it is with the Late Transitional Dialogue, *Theaetetus*, that we
are most interested, since this at once provides clear evidence of
specific progress but also a hint at its limited nature. One of two
contemporary dialogues featuring the then dead Theaetetus (the
other being *Sophist*), they combine to form an eloquent testament
of the high regard in which Theaetetus of Athens (417–369 B.C.E.),
his former student, was held by Plato. With Socrates inevitably as
another main character, the third was Theodorus of Cyrene (465–
398 B.C.E.), himself one-time teacher of both Plato and Theaetetus.

It is in an early passage in *Theaetetus* that incommensurability
reappears and which, somewhat exotically but for good reason,
we give initially in the original Greek:

Περὶ δυνάμεών τι ἡμῖν Θεόδωρος ὅδε ἔγραφε, τῆς τε τρίποδος πέρι καὶ πεντέποδος
[ἀποφαίνων] ὅτι μήκει οὐ σύμμετροι τῇ ποδιαίᾳ, καὶ οὕτω κατὰ μίαν ἑκάστην
προαιρούμενος μέχρι τῆς ἑπτακαιδεκάποδος· ἐν δὲ ταύτῃ πως ἐνέσχετο.

And in Jowett's translation:

> Theodorus was writing out for us something about roots, such
> as the roots of three or five, showing that they are incommen-
> surable by the unit: he selected other examples up to seventeen
> – there he stopped.

The words are those which Plato put into the mouth of Theaetetus
as he discusses with Socrates the nature of knowledge. Theaetetus
recalls to Socrates the memory of Theodorus demonstrating to
him the incommensurability of the (implied) non-square integers
from $\sqrt{3}$ to $\sqrt{17}$.

Evidently, then, progress had been made in those intervening
years, even though its extent is far from clear; most particularly,
there is the implication that Theodorus was in possession of no
general method of establishing incommensurability, otherwise,
why would he repeat his demonstration for each integer?

The final part of the passage is sufficiently intriguing for us to
provide another translation of it[20]:

> ...up to the one of 17 feet; here something stopped him (or:
> here he stopped)...

and one such more[21]

> ...up to seventeen square feet, 'at which point for some reason
> he stopped'.

The Greek, it appears, is ambiguous: did Theodorus stop at, or
before, 17? Whatever the answer to this question, why did he stop
there? And does this curtailment suggest something about the
method he used to establish the incommensurability?

It is natural that these issues have attracted the attention of
scholars, as it is natural that groups of them disagree. Certainly,
by this time the incommensurability of $\sqrt{2}$ was well known and
accepted. The earlier implicit mention in *Meno* is strongly supported by Plato's comment in a letter of rejection to the sponsor
of a student to the academy:

---

[20]From the translation by Arnold Dresden of B. L. van der Waerden, *Science
Awakening* (Groningen, 1954).

[21]T. L. Heath, *A Manual of Greek Mathematics* (London, 1931).

He is unworthy of the name of man who is ignorant of the fact
that the diagonal of a square is incommensurable with its side.

But in the dialogue itself he had Theaetetus continue with[22]:

Now as there are innumerable roots, the notion occurred to us
of attempting to include them all under one name or class.

Which makes it clear that 17 is not an obvious place to stop. The
reason for him stopping at or before 17 has been argued by scholars in sometimes ingenious and inevitably varied ways, including
one of striking pragmatism from G. H. Hardy and E. M. Wright[23]
that

he may well have been quite tired.

There is a particular explanation that has a peculiar intrigue
though, as it appeals to the confidently established fact that, for
Pythagoreans (and Theodorus was a Pythagorean), the parity of
number was of the greatest moment, a point emphasized by the
late mathematician and mathematical historian Bartel van der
Waerden (ibid.):

For the Pythagoreans, even and odd are not only the fundamental concepts of arithmetic, but indeed the basic principles
of all nature.

With their definition of even and odd:

An even number is that which admits of being divided, by one
and the same operation, into the greatest and the least (parts),
greatest in size but least in quantity while an odd number is
that which cannot be so treated, but is divided into two unequal
parts.

The point is that the parity approach that established the irrationality of $\sqrt{2}$, as shown on page 22, has a generalization. It
is described in and may have originated with an elusive article[24] by the amateur German mathematician J. H. Anderhub and

---

[22]Benjamin Jowett's translation.

[23]G. H. Hardy and E. M. Wright, *An Introduction to the Theory of Numbers*,
5th edn (Oxford University Press, 1980).

[24]J. H. Anderhub, Aus den Papieren eines reisenden Kaufmannes (Joco-Seria,
Wiesbaden, Kalle-Werke, 1941).

resurfaced in Wilbur Knorr's unpublished Ph.D. dissertation, later expanded by him to an expensive book,[25] with the historical implications later taken up by Robert L. McCabe.[26]

The observation is that 17 is the first non-square integer of the form $8m + 1$ and for integers of this form the even–odd parity argument fails: that is, Theodorus stopped before 17 and could, using his approach, proceed no further. A form of the reasoning behind the assertion is as follows:

Since any positive integer can be written as one of $4n$, $4n + 1$, $4n + 2$, $4n + 3$ for $n = 0, 1, 2, \ldots$, we can deal with all of them by considering each category separately.

(1) If the integer is of the form $4n$ its square root is $\sqrt{4n} = 2\sqrt{n}$ and we can deal with the case as if it was the number n.

(2) Temporarily omitting the case of $4n + 1$, suppose the integer is of the form $4n + 2$ and write $\sqrt{4n + 2} = a/b$, where a and b are relatively prime. Then $a^2 = (4n + 2)b^2$, which means that a^2 is even and therefore a is even. Write $a = 2k$ to get $a^2 = 4k^2 = (4n + 2)b^2$ and so $(2n + 1)b^2 = 2k^2$ and therefore b^2 is even and so b is even. A contradiction.

(3) Now suppose that the integer is of the form $4n + 3$ and write $\sqrt{4n + 3} = a/b$, where a and b are relatively prime. Then $a^2 = (4n + 3)b^2$. Now, a and b cannot both be even and so at least one of them must be odd and this means that its square is odd; whichever it is, using the previous equation we can conclude that the square of the other must be odd and so itself is odd, therefore, both a and b must be odd. Write $a = 2k + 1$ and $b = 2l + 1$ to get $(2k + 1)^2 = (4n + 3)(2l + 1)^2$, which becomes

$$4k^2 + 4k + 1 = 16nl^2 + 16nl + 4n + 12l^2 + 12l + 3,$$
$$2k^2 + 2k = 8nl^2 + 8nl + 2n + 6l^2 + 6l + 1.$$

With the left-hand side evidently even and the right odd. Again, a contradiction.

---

[25] W. R. Knorr, *The Evolution of the Euclidean Elements: A Study of the Theory of Incommensurable Magnitudes and Its Significance for Early Greek Geometry* (Springer, 1974).

[26] Theodorus' Irrationality Proofs, 1976, *Mathematics Magazine* 49(4):201–3.

(4) Finally, suppose that the integer is of the form $4n + 1$, then either n is even or odd. If n is odd it is of the form $2m + 1$ and so our number will be of the form $4(2m + 1) + 1 = 8m + 5$. We repeat the above arguments to get $\sqrt{8m + 5} = a/b$, where a and b are relatively prime, and so $a^2 = (8m + 5)b^2$ and again this means that both a and b must be odd. Write $a = 2k + 1$ and $b = 2l + 1$ to get $(2k + 1)^2 = (8m + 5)(2l + 1)^2$, which becomes

$$4k^2 + 4k + 1 = 32ml^2 + 32ml + 8m + 20l^2 + 20l + 5,$$
$$k^2 + k = 8ml^2 + 8ml + 2m + 5l^2 + 5l + 1,$$
$$k(k + 1) = 8ml^2 + 8ml + 2m + 5l(l + 1) + 1.$$

Since the product of two consecutive integers must be even this means that the left-hand side is even and the right odd. A contradiction once more.

We are left with the final case that the integer is of the form $4n + 1$ with $n = 2m$ even and so it is of the form $4(2m) + 1 = 8m + 1$ and now the process fails since, if $\sqrt{8m + 1} = a/b$, where a and b are relatively prime, then $a^2 = (8m + 1)b^2$ and again this means that both a and b must be odd. Write $a = 2k + 1$ and $b = 2l + 1$ to get $(2k + 1)^2 = (8m + 1)(2l + 1)^2$, which becomes

$$4k^2 + 4k + 1 = 32ml^2 + 32ml + 8m + 4l^2 + 4l + 1,$$
$$k^2 + k = 8ml^2 + 8ml + 2m + l^2 + l,$$
$$k(k + 1) = 8ml^2 + 8ml + 2m + l(l + 1).$$

And both sides are even: no contradiction this time.

To check, if we attempt an even–odd parity argument with $\sqrt{17}$ we are led nowhere.

Write $\sqrt{17} = a/b$, where a and b are relatively prime, then $a^2 = 17b^2$, which again means that both a and b are odd. Writing $a = 2k + 1$ and $b = 2l + 1$ now yields

$$4k^2 + 4k + 1 = 17(4l^2 + 4l + 1),$$
$$k^2 + k = 17l^2 + 17l + 4,$$
$$k(k + 1) = 17l(l + 1) + 4.$$

There is no contradiction and taking even–odd cases for k and l simply leads to ever more cases.

Whether or not this is the explanation for Theodorus's curtailment of his demonstration we will probably never know, but it is assuredly worth considering as a possibility. In 1942 Anderhub posited a more practical justification, simple, appealing and quite devoid of historical probity: had Theodorus been using a sandbox for his demonstrations, as shown on page 7 of the front matter, he would have overlapped after the hypotenuse of the last triangle is $\sqrt{17}$, making the further demonstration impractical. The figure is constructed as follows. On the hypotenuse of a right-angled isosceles triangle of side 1 is drawn a second right-angled triangle,[27] whose second side is also 1 and the process continued until the hypotenuse of the last triangle is $\sqrt{17}$, where the central angle generated is

$$\sum_{r=1}^{16} \tan^{-1}\left(\frac{1}{\sqrt{r}}\right) = 351.150\ldots°,$$

$$\sum_{r=1}^{17} \tan^{-1}\left(\frac{1}{\sqrt{r}}\right) = 364.783\ldots°.$$

The mathematics of the spiral, the Spiral of Theodorus, associated with the argument (and very much more) was later investigated by Philip J. Davis et al.[28]

The Loss of Similarity

So, the philosophically unacceptable concept of incommensurability had entered the Pythagorean world and through it that of the ancient Greeks in general, with its appearance bringing about the instant demise of the cherished idea that number (positive integers) was the hand-maiden of geometry. The incommensurability of these constructed lengths separated the concepts of arithmetic ratios (which must for them be ratios of positive integers) and the constructible magnitudes of geometry; geometry could cope with the likes of $\sqrt{2}$ in a way that their arithmetic could not. So, the first significant implication of what we now call irrationality was that mathematical enquiry became geometric enquiry, an approach which was to pervade all European mathematics and

---

[27]The marks for right angles and the lengths of sides are omitted for aesthetic reasons.

[28]See Appendix A on page 272.

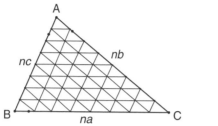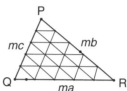

Figure 1.11.

last well into the eighteenth century. A second was that the familiar implications of the comparison of similar figures absconded – and all proofs which rely on them.

To see this, consider what we would call two similar triangles; that is, triangles with corresponding angles equal, and suppose that they are labelled ABC and PQR as shown in figure 1.11. We wish to extract the three corresponding ratios of sides and to do so we have no Pythagorean alternative but to recourse to commensurability. So, suppose that sides BC and QR are commensurable with a common unit a. Suppose further that BC is composed of n of these units and QR is composed of m of them and so divide each line into n and m equal segments, each of length a. In triangle ABC, from the right endpoint of each segment, we draw a line parallel to AB to meet side AC and then repeat this for the corresponding points on AC, drawing lines parallel to the side BC; finally, draw the lines parallel to AC to meet BC. By this means AC will have been divided into n equal segments, say of length b, and similarly AB will have been divided into n equal segments, say of length c, and triangle ABC will have been tessellated by congruent triangles of sides a, b and c. Now move to the similar triangle PQR. By this process it too has been tessellated by these same congruent triangles and this means that we have the proportions of sides as

$$AB:PQ = BC:QR = AC:PR$$
$$(= nc:mc = na:ma = nb:mb = n:m).$$

The ratio of the lengths of the sides of similar triangles is the ratio of two integers: very Pythagorean.

Now let us demonstrate why this makes the escape into geometry simply an escape into another mathematical prison cell.

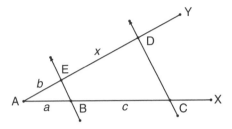

Figure 1.12.

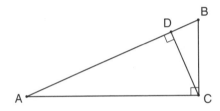

Figure 1.13.

Consider, as one example, their solution of what we would write as the simple linear equation $ax = bc$. The Greek approach is encapsulated in Propositions 8 and 9 in Book VI of *The Elements* but is more easily dealt with by using modern algebraic notation. In figure 1.12 a line AX is drawn with points B and C marked on it distant a and $a+c$ units from A respectively. At some convenient angle to AX a line AY is then drawn with a point E marked on it, distant b units from A; the line BE is then drawn and the line through C parallel to this constructed, intersecting AY at D: the length DE is the value for x.

The modern way of looking at the procedure is to notice that ACD and ABE are similar and so

$$\frac{a}{b} = \frac{a+c}{b+x},$$

which means that $ax = bc$.

And, as a second example, consider figure 1.13. It is a right-angled triangle ABC with the perpendicular from C meeting AB at D, a construction which gives rise to three similar right-angled triangles. Taking the triangles in corresponding order (ABC, ACD and CBD) and using similarity we have

$$\frac{BD}{BC} = \frac{CB}{AB} \quad \text{and} \quad \frac{AD}{AC} = \frac{AC}{AB}$$

and therefore

$$BD = \frac{BC^2}{AB} \quad \text{and} \quad AD = \frac{AC^2}{AB}.$$

Therefore,

$$\frac{BC^2}{AB} + \frac{AC^2}{AB} = BD + AD = AB$$

and so

$$BC^2 + AC^2 = AB^2.$$

As we have already mentioned, it is not known what method (if any) Pythagoras or the Pythagoreans used to prove the famous theorem, but one thing is certain: if they used a simple similarity argument, the discovery of incommensurability would have dealt a devastating blow.

It would take an outstanding insight from one of the greatest thinkers of ancient Greece to deal properly with incommensurable magnitudes and so rescue similarity and thereby the geometric method.

Euclid and Systemization

The influence of *The Elements* in the history of mathematics, in both positive and negative senses, has already been mentioned. Its thirteen "books" constitute the definitive rationalization of the Greek mathematics that had been studied up to its time, both its geometry and also its number theory, and generally there is much confidence in the integrity of the material that has been passed down to us, although the routes it has taken have been long and tortuous.

Unfortunately, such assurance rapidly dissipates when we ask about Euclid himself, about whom almost nothing is known with any certainty. It is true that various ancient authors mention him but without the credibility that is attached to Proclus. We quote again from *The Summary*:

> Not much younger than these [pupils of Plato] is Euclid, who put together the "Elements", arranging in order many of Eudoxus's theorems, perfecting many of Theaetetus's, and also

bringing to irrefutable demonstration the things which had been only loosely proved by his predecessors. This man lived in the time of the first Ptolemy; for Archimedes, who followed closely upon the first Ptolemy makes mention of Euclid, and further they say that Ptolemy once asked him if there were a shorter way to study geometry than *The Elements*, to which he replied that there was no royal road to geometry. He is therefore younger than Plato's circle, but older than Eratosthenes and Archimedes; for these were contemporaries, as Eratosthenes somewhere says. In his aim he was a Platonist, being in sympathy with this philosophy, whence he made the end of the whole "Elements" the construction of the so-called Platonic figures.

To this we may add the conviction that he taught at Ptolemy's great university at Alexandria, that *The Elements* was written around 320 B.C.E. and that mention of him by a modern author adds a little deductive colour[29]:

> In short, it is almost impossible to refute an assertion that *The Elements* is the work of an insufferable pedant and martinet.

And it is to *The Elements* that we must look for evidence of progress in the understanding of irrationality and for that purpose we must concentrate on Books V and X: Book V, the accredited work of Eudoxus of Cnidus (as is Book XII); Book X that of the already mentioned Theaetetus. It is with Euclid's systemization of the work of Eudoxus that we shall start.

If Archimedes of Syracuse (287–212 B.C.E.) was the greatest mathematician of antiquity, surely Eudoxus of Cnidus (408–355 B.C.E.) is runner-up, even though he is primarily acclaimed for his studies in astronomy. None of his original work survives but we do have the supportive witness of (among others) Diogenes Laertius, Proclus and of Archimedes himself, and perhaps the reader will consider it testament enough that two definitions from Books V combine to dispose of the Pythagorean problems with incommensurability in a manner so prescient that the eighteenth-century German mathematician Richard Dedekind was to adapt them in his own definition of irrational numbers, as we shall see in chapter 9. Some of the material which appears is truly remarkable,

---

[29] Salomon Bochner, *The Role of Mathematics in the Rise of Science* (Princeton University Press, 1966).

as it is difficult to understand at a first attempt, a point clearly appreciated by the distinguished nineteenth-century British mathematician Augustus De Morgan in the preface to his own book dedicated to its explanation[30]:

> Geometry cannot proceed very far without arithmetic, and the connexion was first made by Euclid in his Fifth Book, which is so difficult a speculation, that it is either omitted, or not understood by those who read it for the first time. And yet this same book, and the logic of Aristotle, are the two most unobjectionable and unassailable treatises which ever were written.

Our aim is not so grand as to understand the whole book but to demonstrate how Euclid presents the Eudoxian approach to incommensurability and how this approach deals so effectively with the problems brought about by it; to achieve this we will need none of the Book's 25 propositions and just two of its 18 definitions, together with several propositions taken from Books I and VI.

The Axiom of Eudoxus (attributed to him by Archimedes)
Definition 4, Book V

> Magnitudes are said to have a ratio to one another which is capable, when a multiple of either may exceed the other.

Here we should interpret 'is capable' to mean 'makes sense' and initially the definition seems vacuous; surely it is possible for some sufficiently large multiple of any small magnitude to exceed the larger? Not so. First consider an attempted comparison between a finite straight line and an infinite one; no multiple can be found that causes the finite line segment to be increased to exceed the whole line. Second, the definition also precludes attempting to form the ratio of incomparable quantities; for example, a length and an area. Last, and much more subtly, Eudoxus knew of the concept of what has become known as a Horn Angle, that is, the angle between a straight line and a curve, as shown in figure 1.14. Euclid records the concept, if not by name, then by its nature in

[30] *Number and Magnitude. An Attempt to Explain the Fifth Book of Euclid.* Available at http://www.archive.org/details/connexionofnumbe00demorich (accessed 3 October 2011).

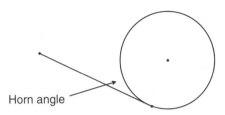

Horn angle

Figure 1.14.

Proposition 16, Book III

The straight line drawn at right angles to the diameter of a circle from its end will fall outside the circle, and into the space between the straight line and the circumference another straight line cannot be interposed, further the angle of the semicircle is greater, *and the remaining angle less, than any acute rectilinear angle.*

The final, italicized phrase has it that a horn angle is less than any rectilinear angle; hence no multiple of the magnitude of a horn angle is greater than that of a rectilinear angle, and again the definition has bite; infinitesimals are thereby obliquely touched upon. In a few lines the teeth of Definition 4 will close to significant effect. With it we have a device which allows comparison between any pair of numbers, commensurables or incommensurables, without ever mentioning the terms and using only integers. For example, the numbers $\sqrt{2}$ and $\sqrt{3}$ may be compared as a ratio since provably $\sqrt{3} > \sqrt{2}$ and $2\sqrt{2} > \sqrt{3}$; similarly, we may compare commensurable with incommensurable numbers: 2 and $\sqrt{2}$ for example; clearly $2 > \sqrt{2}$ and also $2\sqrt{2} > 2$.

So, we have a meaning for the ratio of any two comparable numbers. Now we look to the succeeding definition wherein Eudoxus shows utterly overwhelming insight as he sweeps aside all problems associated with incommensurability – by ignoring it. What follows is his definition of the ratio of any two magnitudes, incommensurable or not; a definition that allowed Greek geometry to move forward after a century of stagnation in the mire of incommensurability:

Definition 5, Book V

Magnitudes are said to be *in the same ratio*, the first to the second and the third to the fourth, when, if any equimultiples whatever are taken of the first and third, and any equimultiples whatever of the second and fourth, the former equimultiples alike exceed, are alike equal to, or alike fall short of, the latter equimultiples respectively taken in corresponding order.

The wording is initially confusing but can be made clearer with the use of a little modern-day algebra, in which case he is saying that (implicitly for magnitudes which can be compared, as in Definition 4 above) the ratio $a:b = c:d$ pertains if for all positive integers m and n:

$$\text{if } na > mb \text{ then } nc > md,$$
$$\text{if } na = mb \text{ then } nc = md,$$
$$\text{if } na < mb \text{ then } nc < md.$$

And this is perhaps made the more reasonable if we build up to it as follows:

For real numbers x and y we can define equality $x = y$ between them by the expedient use of positive integers, arguing that, for all positive integers m and n:

$$\text{if } x > \frac{m}{n} \text{ then } y > \frac{m}{n},$$
$$\text{if } x = \frac{m}{n} \text{ then } y = \frac{m}{n},$$
$$\text{if } x < \frac{m}{n} \text{ then } y < \frac{m}{n}.$$

So, $x = y$ if for all positive integers m and n:

$$\text{if } nx > m \text{ then } ny > m,$$
$$\text{if } nx = m \text{ then } ny = m,$$
$$\text{if } nx < m \text{ then } ny < m.$$

Now write

$$x = \frac{a}{b} \quad \text{and} \quad y = \frac{c}{d}$$

for any *real* numbers a, b, c, d. Then

$$\frac{a}{b} = \frac{c}{d}$$

if for all positive integers m and n:

$$\text{if } n\frac{a}{b} > m \text{ then } n\frac{c}{d} > m,$$

$$\text{if } n\frac{a}{b} = m \text{ then } n\frac{c}{d} = m,$$

$$\text{if } n\frac{a}{b} < m \text{ then } n\frac{c}{d} < m,$$

which of course leads to

$$\text{if } na > mb \text{ then } nc > md,$$
$$\text{if } na = mb \text{ then } nc = md,$$
$$\text{if } na < mb \text{ then } nc < md.$$

Notice that, although it is necessary for a and b to be magnitudes of the same kind and also for c and d to be so, these two kinds need not be the same; for example, a and b might be lengths and c and d areas.

And the Axiom of Eudoxus can be further used here, since it can be shown to render the middle condition above irrelevant, leaving only the inequalities, and further that these can be combined in the following satisfying way[31]:

Two ratios are equal when no rational fraction whatever lies between them.

Let us briefly compare the two types of proof, the one using commensurability, the other the approach of Eudoxus. To do so consider:

Proposition 1, Book VI

Triangles and parallelograms which are under the same height are to one another as their bases.

If we concentrate on triangles, this states that the ratio of the areas of two triangles of the same height is the ratio of their bases. Both proofs rely on an earlier proposition, not affected by commensurability issues.

---

[31]Otto Stolz, 1885, *Vorlesungen über allgemeine arithmetik: nach den neueren ansichten* (Cornell University Library).

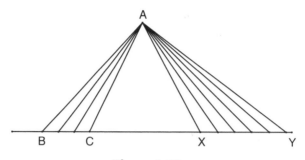

Figure 1.15.

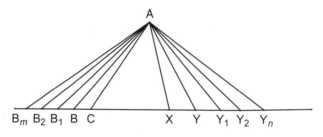

Figure 1.16.

Proposition 38, Book I

Triangles which are on equal bases and in the same parallels equal one another.

That is, triangles having the same base and height are of equal area.

First, let us consider a proof of Proposition 1 based on commensurability.

If the triangles are ABC and AXY, as in figure 1.15, since BC and XY are commensurable there is a common unit u which measures them both; suppose that $BC = nu$ and that $XY = mu$. On each base mark off the n and m equal length segments respectively and join to A to form n and m triangles respectively, each having a common altitude and equal base. Then $\triangle ABC$ has been divided into n triangles each of equal area and $\triangle AXY$ into m triangles each equal to that same area. Therefore, $\triangle ABC : \triangle AXY = n : m = BC : XY$.

And now consider one where commensurability is not assumed.

Let the two triangles in question be once again ABC and AXY. Referring to figure 1.16, on CB extended, for arbitrary positive integers m and n, mark off m equal segments and connect to A: similarly on XY extended mark off n equal segments and connect to A. Then $B_{m-1}C = m\mathrm{BC}$ and $\Delta AB_{m-1}C = m\Delta ABC$ and $XY_n = n\mathrm{XY}$ and $\Delta AXY_n = n\Delta AXY$. If we adopt the symbol $>=<$ to summarize the triple Eudoxian condition, we have $\Delta AB_mC \; >=< \; \Delta AXY_n$ according as $B_mC \; >=< \; XY_n$ and $m\Delta ABC \; >=< \; n\Delta AXY$ according as $m\mathrm{BC} \; >=< \; n\mathrm{XY}$ for all positive integers m and n.

Therefore $\Delta ABC : \Delta AXY = \mathrm{BC} : \mathrm{XY}$ as the proposition required. No mention of commensurability of lengths: no integer ratios.

Finally, if we are to gauge the importance of the impact of incommensurability, with its concomitant that any result depending on similarity was doomed, we should look to Pythagoras's Theorem itself. We have seen the earlier, neat derivation, which does rely on similarity, but in *The Elements* Euclid has the result appear (as ever, without attribution) as Proposition 47 of Book 1, with his systematization requiring it for a number of other propositions throughout the work, starting in Book 2; all of this long before the Eudoxian approach of Book V. He needed it, he couldn't prove it using any ideas of similarity and so he provided the famous *bride's chair* argument, which has been attributed (by Proclus and others) to Euclid himself. We take the opportunity remind the reader of its comparatively involved detail.

Proposition 47, Book 1

> In right-angled triangles the square on the side opposite the right angle equals the sum of the squares on the sides containing the right angle.

The *bride's chair* is shown in figure 1.17, with the argument proceeding as follows.

Describe the square BDEC on BC, and the squares BFGA and AHKC on BA and AC. Draw AL through A parallel to BD, and join AD and FC.

Since each of the angles BAC and BAG is a right angle, CA is in a straight line with AG.

For the same reason BA is also in a straight line with AH.

The angle DBA equals the angle FBC, since each has angle ABC common with the remainder of each a right angle.

Since BD equals BC, and BF equals BA, the two sides AB and BD equal the two sides FB and BC respectively, and the angle ABD

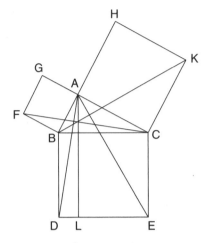

Figure 1.17.

equals the angle FBC, therefore the base AD equals the base FC, and the triangle ABD equals the triangle FBC.

Now the rectangle BL is double the triangle ABD, for they have the same base BD and are in the same parallels BD and AL. And the square BFGA is double the triangle FBC, for they again have the same base FB and are in the same parallels FB and GC.

Therefore the rectangle BL also equals the square BFGA.

Similarly, if AE and BK are joined, the rectangle CL can also be proved equal to the square ACKH. Therefore the whole square BDEC equals the sum of the two squares BFGA and ACKH.

And the square BDEC is described on BC, and the squares BFGA and ACKH on BA and AC.

Therefore the square on BC equals the sum of the squares on BA and AC.

Therefore in right-angled triangles the square on the side opposite the right angle equals the sum of the squares on the sides containing the right angle.

Clever, but in terms of effort involved, a heavy price to pay to have the result available before its time; if we look to Book V1, the result is generalized as Proposition 31 by having rectangles on each of the sides of the right-angled triangle – and the proof *is* based on similarity!

With incommensurability of magnitudes acknowledged and tamed, we move to Book X, by far the longest (it constitutes about

one quarter of *The Elements*) and the most challenging of the thirteen books, and the one devoted to summarizing the arithmetic
of incommensurables.

Reactions to the work are varied, with Augustus De Morgan saying of it that 'this book has a completeness which none of the others (not even the fifth) can boast of', Sir Thomas Heath that Book X
'is perhaps the most remarkable, as it is the most perfect in form,
of all the Books of *The Elements*', to be balanced by the view of
the one of the twentieth-century's most distinguished historians
of ancient mathematics, the late and greatly lamented Stanford
academic, Wilbur Knorr (who also contributed to the study of the
Spiral of Theodorus[32]):

> The student who approaches Euclid's Book X in the hope that
> its length and obscurity conceal mathematical treasures is
> likely to be disappointed. As we have seen, the mathemati-
> cal ideas are few and capable of far more perspicuous expo-
> sition than is given them here. The true merit of Book X, and I
> believe it is no small one, lies in its being a unique specimen of
> a fully elaborated deductive system of the sort that the ancient
> philosophies of mathematics consistently prized. It constitutes
> the results of a detailed academic exercise to codify the forms
> of the solutions of a specific geometric problem and to demon-
> strate a basic set of properties of the lines determined in these
> solutions. One can thus profitably study Book X to learn how
> its author sought to convert a body of geometric findings into
> a system of mathematical knowledge.

It is fortunate that scholars have access to a reliable copy (originally in Arabic) of the *Commentary of Pappus* on the work and it
is with his authority[33] that the name of Theaetetus is associated
with it.

Our purposes will be served by noting the underpinning first
four definitions and following them with the first few of a long
sequence of propositions:

Definition 1, Book X

---

[32]Wilbur Knorr, 1983, La Croix des Mathématiciens: The Euclidean Theory of
Irrational Lines, *Bulletin of the American Mathematical Society* 9(1).

[33] *The Commentary of Pappus on Book X of Euclid's Elements*, translation from
the Arabic by William Thomson (Harvard University Press, 1930).

Those magnitudes are said to be *commensurable* which are measured by the same measure, and those *incommensurable* which cannot have any common measure.

The inevitability of the existence of incommensurable magnitudes is acknowledged.

Definition 2, Book X

Straight lines are *commensurable in square* when the squares on them are measured by the same area, and *incommensurable in square* when the squares on them cannot possibly have any area as a common measure.

The distinction is made between two types of incommensurables: those which, although incommensurable in themselves, are commensurable in square and those which are not. The notion distinguishes, for example, between $\sqrt{2}$ and $\varphi = \frac{1}{2}(1 + \sqrt{5})$; $\sqrt{2}$ is commensurable in square with the unit since $(\sqrt{2})^2 = 2$ but φ not so since $\varphi^2 = \varphi + 1$ (see page 5).

Definition 3, Book X

With these hypotheses, it is proved that there exist straight lines infinite in multitude which are commensurable and incommensurable respectively, some in length only, and others in square also, with an assigned straight line. Let then the assigned straight line be called *rational*, and those straight lines which are commensurable with it, whether in length and in square, or in square only, *rational*, but those that are incommensurable with it *irrational*.

Definition 4, Book X

And let the square on the assigned straight line be called *rational*, and those areas which are commensurable with it *rational*, but those which are incommensurable with it *irrational*, and the straight lines which produce them *irrational*, that is, in case the areas are squares, the sides themselves, but in case they are any other rectilineal figures, the straight lines on which are described squares equal to them.

With these definitions we see that both incommensurables and commensurables are each infinite in number, if a little confusing. Next, another contribution from Eudoxus:

Eudoxus's Method of Exhaustion
Proposition 1, Book X

> Two unequal magnitudes being set out, if from the greater
> there is subtracted a magnitude greater than its half, and from
> that which is left a magnitude greater than its half, and if this
> process is repeated continually, then there will be left some
> magnitude less than the lesser magnitude set out.

Rather than reproduce the original proof from *The Elements*, we
will demonstrate an equivalent, which demonstrates another use
of the Axiom of Eudoxus.

Suppose that we have the greater magnitude AB and the lesser
CD, then the axiom tells us that there is a positive integer m such
that $mCD > AB$, which means that $CD : AB > 1 : m$. It is also clear
that there is a positive integer n such that $2^n > m$. Therefore there
is an n such that $CD : AB > 1 : 2^n$, which means that $(1/2^n)AB <$
CD, with the left-hand side of this inequality an upper bound for
the length remaining after the nth cut-off of AB. Therefore, that
magnitude is less than CD, as required.

Again, the force of the result may not be apparent at first, but
that magnitude CD can be chosen to be as small as we wish; in
modern notation we might well choose the letter ε to represent it
and with that choice we are easily led to a modern refrain:

> ...given $\varepsilon > 0$ there exists an integer n so that {such-and-such
> a quantity depending on n} is less that ε...

and we have a limiting process which, in the right hands, can
be put to extraordinary use. For example, in a letter to his fre-
quent correspondent Eratosthenes, Archimedes himself records
that, using this technique, Eudoxus was the first to prove that the
volume of the cone and the pyramid are one-third respectively of
the cylinder and prism with the same base and height; a result that
was known to Democritus but one which he could not establish in
a rigorous way. Archimedes was to develop the technique to bring
about a host of important and impressive results of quadrature,
presaging integral calculus by about 2000 years.

Now we turn to the subsequent proposition, which gives a criterion for incommensurability, although its general usefulness is
open to question: when does 'never' happen?

Proposition 2

If, when the lesser of two unequal magnitudes is continually subtracted in turn from the greater that which is left never measures the one before it, then the two magnitudes are incommensurable.

And then to a short sequence of propositions which characterize commensurability:

Proposition 5

Commensurable magnitudes have to one another the ratio which a number has to a number.

Proposition 6

If two magnitudes have to one another the ratio which a number has to a number, then the magnitudes are commensurable.

Proposition 7

Incommensurable magnitudes do not have to one another the ratio which a number has to a number.

Proposition 8

If two magnitudes do not have to one another the ratio which a number has to a number, then the magnitudes are incommensurable.

With commensurability so defined, later we find an initially bewildering list of differing incommensurables, generated from two given magnitudes a and b which in themselves are incommensurable but which are commensurable in square: their *medial* is the *mean proportion* (geometric mean) $\sqrt{ab}$, their *binomial* $a+b$ and their *apotome* $a - b$ (with $a > b$). Once again, referring to Pappus

Theaetetus distinguished the powers commensurable in length from those incommensurable, and he distributed the very well known among the surd lines according to the means, so that he assigned the medial line to geometry and the binomial to arithmetic and the apotome to harmonics, as Eudemus the Peripatetic reported.

As the dialogue develops Euclid methodically addresses all possible magnitudes which we would write in the form

$$\sqrt{\sqrt{a} \pm \sqrt{b}}.$$

And this is where the opening chapter in the development of irrationality should end: there is much more mention of them and work done with them and on them, but nothing of great consequence. Incommensurables had been discovered and geometry seen as the place to deal with them, with arithmetic firmly anchored in the commensurable – which is where matters stood for many centuries. The irrational number had yet to be born.

The Route to Germany

> One of the endlessly alluring aspects of mathematics is that its thorniest paradoxes have a way of blooming into beautiful theories.
>
> Philip J. Davis

In Chapter 1 we considered the efforts of the ancients as they struggled to come to terms with incommensurability. Here we will move through time and place at some speed as we continue the story from where it was left, to move with it through ancient India and Arabia, medieval Europe and finally nineteenth-century Germany.

The ancient Greeks had not embraced irrationality, they had avoided it as much as possible and used geometry to best cope with it, and the succession of their civilization by that of the Romans did nothing to advance matters. What did the Romans do for us? According to the satirical (and hilarious) Monty Python film *The Life of Brian* (1979), the dissident Reg asked the same question, which brought about the following dialogue:

> REG. All right, but apart from the sanitation, medicine, education, wine, public order, irrigation, roads, the fresh water system and public health, what have the Romans ever done for us?
> ATTENDEE. Brought peace?
> REG. Oh, peace – shut up!

And of course they brought us the Julian calendar. Yet nothing in Reg's list, and any that are more comprehensive, wanted for the integrity of mathematical process, much less that some measure is incommensurable with the unit. Theirs was the mantra of pragmatism: for the pure mathematician the Romans did nothing. It is enough to realize that, in the whole period of their empire (which lasted from about 750 B.C.E. to 476 C.E., when the last Roman emperor was deposed) there is not one recorded Roman mathematician of note. On the contrary, Morris Kline[1] would have it that

> their entire role in the history of mathematics was that of an agent of destruction.

Surely this is an extreme view and, focused on pure mathematics as it is, ignores their gigantic contributions, as suggested in the above dialogue, and which embraced the intellectual spectrum from philosophy to engineering. Yet, if it is news of advances with irrational numbers that we seek, we must look further afield, to find the mathematical successors to the Greeks in the Indian and Arabian civilizations.

Hindus and Arabs

The Hindu civilization dates back thousands of years B.C.E. but it is the *high period* of this ancient people, which dates from about 200 C.E. to 1200 C.E., that brought with it considerable advances in mathematical thought and procedure. Their motivation for mathematical enquiry was not the desire for the abstract purity of the Greeks, but a mixture of the need to cope with the practical (accounting, astronomy and astrology) and the desire to understand the theoretical (in particular, the number system). Theirs was the first acceptance of 0 and of negative numbers, it was they who first introduced the base 10 positional number system and it was they who first manipulated square roots with some degree of conviction. It is not that this evidenced philosophical progress with irrationals, just that they were accepted as numbers and manipulated in the same manner as rationals: the geometric *incommensurable* became the arithmetic *irrational*.

Let us see[2] how one of the leading astronomers and mathematicians of the time, Śrīpati (1019–1066 C.E.), approached

---

[1] Morris Kline, *Mathematical Thought from Ancient to Modern Times* (Oxford University Press, 1972), p. 178.

[2] Bibhutibhusan Datta and Awadhesh Narayan Singh, 1993, Surds in Hindu mathematics, *Indian Journal of History of Science* 28(3):254.

the problem in his work on astronomy and arithmetic entitled *Siddhantasekhara*:

> For addition or subtraction, the surds[3] should be multiplied (by an optional number) intelligently (selected), so that they become squares. The square of the sum, or difference of their roots, should then be divided by that optional multiplier. Those surds which do not become squares on multiplication (by an optional number), should be put together (side by side).

For the ancient Hindus a surd did not involve a square root but was a non-square positive integer. For two such integers a and b his prescription results in

$$\frac{1}{c}(\sqrt{ac} \pm \sqrt{bc})^2 = (\sqrt{a} \pm \sqrt{b})^2$$

and so

$$\sqrt{a} \pm \sqrt{b} = \sqrt{\frac{1}{c}(\sqrt{ac} \pm \sqrt{bc})^2}.$$

Of course, the idea is to create perfect squares by the introduction of a factor, which is only possible if all prime factors of a and b appear with the same parity of power. If we take $a = 3$, $b = 12$, we can take $c = 12$ to yield

$$\sqrt{3} + \sqrt{12} = \sqrt{\frac{1}{12}(\sqrt{36} + \sqrt{144})^2} = \sqrt{\frac{1}{12}(6 + 12)^2} = \sqrt{\frac{18^2}{12}}$$
$$= \sqrt{27} = 3\sqrt{3}.$$

Cumbersome to our modern eye, but assuredly a perceptive idea.

A little later we have a contribution from one of the leading mathematicians of the twelfth century, Bhaskara II (1114–1185 C.E.), from the *Bijaganita*:

> Suppose the sum of the two numbers of the surds as the greater surd and twice the square root of their product as the lesser. The addition or subtraction of these like integers is so.

---

[3]The translator's choice of word. We will deal with the use of the word *surd* a little later.

That is, the integer identity $a + b \pm 2\sqrt{ab} = (\sqrt{a} \pm \sqrt{b})^2$ means that

$$\sqrt{a} + \sqrt{b} = \sqrt{a + b + 2\sqrt{a}\sqrt{b}} = \sqrt{a + b + 2\sqrt{ab}}.$$

Again with our (and his) example,

$$\sqrt{3} + \sqrt{12} = \sqrt{3 + 12 + 2\sqrt{3}\sqrt{12}} = \sqrt{3 + 12 + 2\sqrt{36}}$$
$$= \sqrt{15 + 12} = \sqrt{27} = 3\sqrt{3}.$$

Other equivalent prescriptions for surd addition and subtraction appeared throughout Hindu mathematics:

$$\sqrt{a} \pm \sqrt{b} = \sqrt{b\left(\sqrt{\frac{a}{b}} \pm 1\right)^2},$$

$$\sqrt{a} \pm \sqrt{b} = \sqrt{\frac{1}{a}(a \pm \sqrt{ab})^2},$$

$$\sqrt{a} \pm \sqrt{b} = \sqrt{c\left(\sqrt{\frac{a}{c}} \pm \sqrt{\frac{b}{c}}\right)^2}.$$

Following these, instructions for the multiplication and division of expressions involving sums and differences of surds are variously listed, with Bhaskara II demonstrating that

$$\frac{\sqrt{9} + \sqrt{450} + \sqrt{75} + \sqrt{54}}{\sqrt{25} + \sqrt{3}} = \sqrt{18} + \sqrt{3}$$

by a method too harrowing to dwell upon. And if that is not enough, he also demonstrated that the square root of $16 + \sqrt{120} + \sqrt{72} + \sqrt{60} + \sqrt{48} + \sqrt{40} + \sqrt{24}$ is $\sqrt{6} + \sqrt{5} + \sqrt{3} + \sqrt{2}$ and gives a fulsome explanation of a criterion which can be applied to judge whether multinomial surds have such square roots, pointing out that "This method has not been explained at length by previous writers. I do it for the instruction of the dull."

In their own way the Hindus could manipulate irrationals of the form that is the square root of non-square integers. They had trigonometry too and, using their base 10 number system, tables to calculate values of trigonometric functions to impressive accuracy; Madhava of Sangamagramma (1350–1425 C.E.) founded analysis when he gave infinite series for the trigonometric functions (predating Taylor and Maclaurin by several centuries) and through them, the value for π correct to 11 decimal places. Their contribution to mathematics as a whole and to the manipulation of surd irrationals in particular helped to bridge the gap between the mathematics of ancient Greece and that of the European Renaissance; but the Arabs contributed too.

Caliph Omar ibn Khattab (ca. 580–644 C.E.), architect and second ruler of the Islamic caliphate, began what was to be a vast expansion of the Islamic empire. Attached to him is a tradition, persistent but scant on verification, that it was he who was responsible for the final destruction of the great Library of Alexandria, condemning the thousands of manuscripts to the flames with the epigram:

> Either the books contain what is in the Koran, in which case we
> do not have to read them, or they contain the opposite of what
> is in the Koran, in which case we must not read them.[4]

Modern scholarship casts severe doubt on the matter but none at all on the importance of his caliphate. The expansion of the Arabic empire was to continue so that, by the eighth century C.E., it extended from North Africa to southern Europe and through the Middle East to the western edges of India. Contact with India ensured transmission of the many mathematical achievements of the Hindus, including the decimal number system, and contact in general brought the major works of the Greeks to the new imperial capital of Baghdad. Caliph al-Mansur had founded his capital in 762 C.E. and his reign heralded the time of universal scholarship, most particularly with that of the previously suppressed Persians; a successor, Caliph Harun al-Rashid, arranged for the first translations of many Greek mathematical texts into Arabic (including Euclid's *Elements*) and on his own succession his son, Caliph al-Ma'mūn, went further and founded the *Bayt al-Hikma* (House of Wisdom), a form of academy of science which was to last more than 200 years. Here were gathered the learned manuscripts in Greek and Sanskrit and scholars who could translate them;

---

[4]To be compared with the alleged comment of St Augustine (354–430 C.E.): "Whatever knowledge man has acquired outside of Holy Writ, if it be harmful it is there condemned; if it be wholesome it is there contained."

over the years many translations of important Greek and Indian mathematical books were added to that of *The Elements*.

One of the thinkers and translators at work in the House of Wisdom was one Abu Ja'far Muhammad ibn Mūsā Al-Khwārizmī (ca. 780 C.E. to ca. 850 C.E.), or Al-Khwārizmī for short. We owe him much as a thinker but also we are indebted to him for some of the words that find common use in mathematics: *algorithm* and *algebra*, for example. It is also probable that we must credit him with influencing the inclusion of *surd* in the mathematical lexicon, in the following manner.

We can look to Plato's earlier Dialogues, the *Hippias Major* (if indeed he was its author) and *The Republic*, as evidence that, at that stage, incommensurables were referred to in Greek as $\alpha\rho\rho\epsilon\tau o\varsigma$, which translates into Latin as *arrhetos* (or *alogos*), which mean *unmentionable* or *inexpressible*. Al-Khwārizmī's Arabic translation of this was to *asamm* or *deaf* and when this was subsequently translated into Latin, possibly by Gherardo of Cremona, it was to *surdus*, *surde* or *surd*, meaning *mute* or *deaf*. We should also mention that there is no common ground to the modern meaning of the word: $\sqrt{2}$ is assuredly a surd, but are $1 + \sqrt{2}$ and $\sqrt{\pi}$, for example?

It was the development of algebra that was to progress the use of irrational numbers significantly in that it blurred the distinction between types of number: integer, rational or irrational could be treated as a root of a particular type of equation; manipulate the equation and you manipulate the number – whatever it is. And the Arabs were most adept in its use, even though their approach was rhetorical and without the use of the symbolism that we now associate with the discipline.

A second significant (and equally challenging) name is that of Al-Khwārizmī's intellectual successor, Abu Kamil Shuja ibn Aslam ibn Muhammad ibn Shuja (ca. 850 C.E. to ca. 930 C.E.), or Abu Kamil. Of his life precious little is certain, but of his work we know more – and enough to judge that he was gifted with a remarkable facility for manipulating extremely complex surds. He was happy to work with them as numbers, whether they be the roots of some equation or, indeed, coefficients in an equation.

We will consider just two examples of his work, with each amply justifying his epithet of *the Egyptian calculator*.

From his work *The Pentagon and the Decagon*, Example XVI
has him finding the diameter of the circumscribing circle of a
pentagon of side 10, which he generalizes by the instruction[5]:

1. It is obvious that you multiply one of its sides by itself then
   you double this and keep it.
2. Then multiply again one of its sides by itself.
3. Then what is kept by itself.
4. Then take 2/5 of it whose square root you find (by adding).
5. Then add the results to what you kept.
6. Then take the square root of the sum.
7. What is left is the diameter of the circle.

With a pentagon of side a the instructions becomes in modern
notation:

1. $a^2 \to 2a^2$.
2. $a^2 \to (a^2)^2$.
3. $2a^2 \to (2a^2)^2$.
4. $\frac{2}{5}\sqrt{(2a^2)^2 + (a^2)^2}$.
5. $2a^2 + \frac{2}{5}\sqrt{(2a^2)^2 + (a^2)^2}$.
6. $\sqrt{2a^2 + \frac{2}{5}\sqrt{(2a^2)^2 + (a^2)^2}}$.

The reader may wish to perform a little algebra to show that this
does give the standard result of $\frac{1}{5}a\sqrt{50 + 10\sqrt{5}}$ – and then prove
it.

The second example is from his surviving work, *Algebra*, which
was at once a commentary on and an extension to the second
of Al-Khwārizmī's publications above. Al-Khwārizmī's forty problems were extended to sixty-nine, an extension not only of variety but also difficulty: we will take problem 61 as an indicator
of the manner in which surds were now being manipulated. It is
wise to spare ourselves the loquacious original statement of the
problem, much less the description of its solution; as we have
just mentioned, Arabic mathematics was at that time expressed
entirely verbally. In modern form it asks for the triplet satisfying

[5]Mohammad Yadegari and Martin Levey, 1971, Abu Kamil's 'On the Pentagon and the Decagon', *History of the Science Society of Japan* (Tokyo)
Supplement 2, p. 1.

the simultaneous equations

$$x + y + z = 10,$$
$$xz = y^2,$$
$$x^2 + y^2 = z^2.$$

His method was that of 'false position', a technique which had been in use long before these times but which Abu Kamil took to giddy heights; the principle is to make a first guess of the solution and to correct matters afterwards. Here he took his starting point as $x = 1$. This causes the second pair of equations to become $z = y^2$ and $1 + y^2 = z^2$ to yield the bi-quadratic equation $1 + y^2 = y^4$ the positive solution of which he wrote as

$$y = \sqrt{\tfrac{1}{2} + \sqrt{1\tfrac{1}{4}}} \quad \text{and so} \quad z = \tfrac{1}{2} + \sqrt{1\tfrac{1}{4}}.$$

Now notice that with the last two equations, if we have a triplet of solutions, then any multiple of the triplet will also be a solution. This means that any multiple

$$k\left(1, \sqrt{\tfrac{1}{2} + \sqrt{1\tfrac{1}{4}}}, \tfrac{1}{2} + \sqrt{1\tfrac{1}{4}}\right)$$

of this initial solution set also satisfies them. Now he found the value of k that ensures the first equation is satisfied by writing

$$k\left(1 + \sqrt{\tfrac{1}{2} + \sqrt{1\tfrac{1}{4}}} + \tfrac{1}{2} + \sqrt{1\tfrac{1}{4}}\right) = 10$$

or

$$k\left(1\tfrac{1}{2} + \sqrt{\tfrac{1}{2} + \sqrt{1\tfrac{1}{4}}} + \sqrt{1\tfrac{1}{4}}\right) = 10.$$

The triplet

$$\frac{10}{1\tfrac{1}{2} + \sqrt{\tfrac{1}{2} + \sqrt{1\tfrac{1}{4}}} + \sqrt{1\tfrac{1}{4}}}\left(1, \sqrt{\tfrac{1}{2} + \sqrt{1\tfrac{1}{4}}}, \tfrac{1}{2} + \sqrt{1\tfrac{1}{4}}\right)$$

is then a solution to the set of equations.

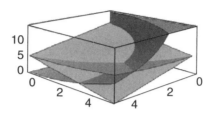

Figure 2.1.

And he simplified them to a form which we would write as

$$x = \tfrac{5}{4}\left(4 - 3\sqrt{2(1 + \sqrt{5})} + \sqrt{10(1 + \sqrt{5})}\right) = 2.57066\dots,$$

$$y = \tfrac{5}{2}\left(1 - \sqrt{5} + \sqrt{2(1 + \sqrt{5})}\right) = 3.26993\dots,$$

$$z = \tfrac{5}{4}\left(2(1 + \sqrt{5}) + \sqrt{2(1 + \sqrt{5})} - \sqrt{10(1 + \sqrt{5})}\right) = 4.15941\dots,$$

with Abu Kamil having found the single point of intersection of the three surfaces in figure 2.1.

With this scant appreciation of Abu Kamil's work we will leave him but with the thought that *Algebra* (in particular) was the basis of the mathematical work of the much travelled Fibonacci of Pisa, whom we will discuss in a few lines, and as such brought the discipline of algebraic formulation and manipulation of problems to Europe – unfortunately in the unenlightened Middle Ages.

And we pass another great scholar in Abū Bakr ibn Muhammad ibn al Husayn al-Karajī, or al-Karajī (953 to ca. 1029 C.E.); engineer and mathematician, and another source for Fibonacci. His geometric proof that

$$1^3 + 2^3 + 3^3 + \cdots + 10^3 = (1 + 2 + 3 + \cdots + 10)^2$$

stands as a model of ingenuity but his major contribution to the history of incommensurability was to suggest the replacement of Euclid's geometric irrational magnitudes with irrational numbers. In chapter 6 of his *Wonderful Book on Hisāb*,[6] written in Baghdad around the year 1000 C.E., he discussed the medials, binomials and apotomes of Euclid, commenting[7]

---

[6]Meaning, *calculation* (including algebraic manipulation).

[7]J. Lennart Berggren, *Mathematics in Medieval Islam*. Part of *The Mathematics of Egypt, Mesopotamia, China, India and Islam. A Sourcebook*, ed. V. J. Katz (Princeton University Press, 2007).

> Now I shall show you how these terms can be transferred to numbers and shall add to them because they are insufficient for Hisāb, due to its extent.

As we scamper through names complex to the Western eye we should mention Ghiyath al-Din Abu'l-Fath Umar ibn Ibrahim Al-Nisaburi al-Khayyami, or Omar Khayyam (1048–1122 C.E.). Renowned equally for his poetry and mathematics, he too was happy with the idea of incommensurable magnitudes being thought of as numbers and most particularly, in his most famous mathematical work, *Treatise on Demonstration of Problems of Algebra*, he took the philosophically vast leap of developing a numerical version of Eudoxus's theory of proportion that comes very close to the rigorous definition of irrational numbers put forward by Richard Dedekind in the nineteenth century, to which we have already alluded and which in this chapter we approach. His prescience is further emphasized when, in an early paper,[8] he discussed a geometric problem which led him to the cubic equation which we would write as

$$x^3 + 200x = 20x^2 + 2000.$$

He solved the equation numerically by interpolation and the use of trigonometric tables, and geometrically by the intersection of a hyperbola and a circle and made the astonishingly insightful comment:

> This cannot be solved by plane geometry, since it has a cube in it. For the solution we need conic sections.

We will consider that particular piece of late nineteenth-century mathematics in chapter 7, finishing our fleeting visit to this great scholar with the celebrated Edward Fitzgerald translation of the most famous quatrain from his *Rubaiyat*, which might well be entitled *An Author's Lament*:

> The Moving Finger writes, and, having writ,
> Moves on: nor all thy Piety nor Wit
> Shall lure it back to cancel half a Line,
> Nor all thy Tears wash out a Word of it.

---

[8]Omar Khayyam, 1963, A paper of Omar Khayyam, *Scripta Mathematica* 26:323–37.

With this necessarily but frustratingly scant visit to the imaginative and vastly influential world of the Hindus and Arabs we must move to the Europe of the Middle Ages for further news of irrational numbers – and unavoidably to Fibonacci of Pisa.

Fibonacci

The (variously named) northern Algerian city of Béjaïa (in Italian, Bugia) was part of that vast twelfth-century Muslim empire and an important port, cultural and trading centre. The local interests of the merchants of the Italian city state of Pisa were for a period represented there by one Guglielmo Bonacci and it is not surprising, then, that the city became the place of schooling of Guglielmo's son, Leonardo; Leonardo Bonacci, Leonardo Bigollo, Leonardo Pisano or, most recognizably to the modern eye, Fibonacci of Pisa[9] (1170–1250 C.E.), his posthumous nickname. The noun *Fibonacci* is now almost invariably taken as an adjective, qualifying its own noun, *sequence*, which arises from the solution to a rabbit-breeding problem proposed by him; he was much more though and is widely regarded as the greatest number theorist in the 1300 year period between Diophantus in the fourth century C.E. and Fermat in the seventeenth century. The boy learned about the Hindu–Arabic decimal number system and later the man was to undertake extensive travels in the Middle East and garner more mathematical knowledge, unknown in the Europe of the Dark Ages. A significant distillation of his accumulated knowledge is his extensive work on algebra, the encyclopaedic *Liber Abaci* (*The Book of Calculation*), which first appeared in 1202, with a second edition in 1228. Here is found the details of that number system, with its numerals 1, 2, 3, 4, 5, 6, 7, 8, 9 and with its *zephirum*, 0, (the Latinized Arabic word *zephr*), which was to become *zefiro* in Italian and *zero* in the Venetian dialect): for the first time, a readable exposition for the non-specialist was available which explained the great benefits of replacing the still-used Roman (virtually) non-positional number system with one that lent itself to practical arithmetic problems: the fourth section of the book has him giving numeric approximations of surds, thereby removing them entirely from geometry. This said, he was given to working

---

[9]From *filius Bonacci*, or son of Bonaccio.

in *sexagesimal*, or base 60. It is a sadness that other works by him have not survived, most particularly a commentary on Book X of *The Elements*, which he is known to have written and which we know contained a full numeric treatment of Euclid's incommensurable numbers. Another of his works has survived, though: *Flos* (*Flower*). Written in 1225, this contrastingly slim volume comprises the statement of and answers to three mathematical challenges presented to Fibonacci by one Johannes of Palermo; one of them is of interest to us. Fibonacci's fame as mathematician and expositor was by this time well established and it is small surprise that he came to the attention of that great patron of science and of learning in general, the *Stupor Mundi*[10] or Holy Roman Emperor, Frederick II. Frederick's court may have been "a picturesque Arabian Nights court at Palermo, enlivened by dancing girls, jugglers, musicians, eunuchs, and an exotic menagerie"[11] but it was also populated with great thinkers both from the arts and the sciences, one of whom was the philosopher and translator Johannes of Palermo. Although based in Palermo, the court was peripatetic and in 1225 found its way to that great city of Pisa, and Frederick desired to meet Europe's leading mathematician who was introduced by Fibonacci's friend, the imperial astronomer Dominicus Hispanus. Johannes was another friend and, as was customary, offered intellectual challenges to the revered guest, in this case by posing three problems, the second of which was to find the (only real) root of the equation

$$x^3 + 2x^2 + 10x = 20.$$

His arguments quickly establish that such a root cannot be integral or rational, moreover, that it could not take any of the forms from Book X of Euclid's *Elements*, and as such it represented a new type of irrational number, one that is not capable of construction by straight edge and compass, echoing Omar Khayyam. He continued with the comment[12]:

---

[10]Wonder of the world.

[11]Joseph and Frances Gies, 1969, *Leonard of Pisa and the New Mathematics of the Middle Ages* (Thomas Y. Crowell).

[12]B. Boncompagni, transl., 1857–1862, *Fibonacci's Flos*, in *Scritti di Leonardo Pisano: mathematico del secolo decimoterzo, Tipografia delle Scienze Mathematiche e Fisiche* 2:227–53.

And because it was not possible to solve this equation in any other of the above ways, I worked to reduce the solution to an approximation

and then, without justification, states that the root is approximately $1; 22, 07, 42, 33, 04, 40$ in *sexagesimal* notation. In decimal form this becomes

$$1; 22, 07, 42, 33, 04, 40 = 1 + \frac{22}{60} + \frac{7}{60^2} + \frac{42}{60^3} + \frac{33}{60^4} + \frac{4}{60^5} + \frac{40}{60^6}$$
$$= 1.36880810785$$

a rational approximation to nine decimal places of accuracy of the irrational number

$$\frac{1}{3}\left(-2 - \frac{13 \times 2^{2/3}}{(176 + 3\sqrt{3930})^{1/3}} + (352 + 6\sqrt{3930})^{1/3} \right).$$

Irrational numbers were ever more frequently appearing in texts, with the hint of their variety becoming ever stronger. We shall look at non-constructible numbers on page 203 but now we move to an example of the acceptance of irrationality – and more than that, an early and fantastic use to which it could be put.

The Perfect Year

History has treated the memory of the Scholastic (John) Duns Scotus[13] (ca. 1266–1308 C.E.) somewhat variously. His delicate philosophical and theological arguments had him accorded the sobriquet *the Subtle Doctor*; his treatises on grammar, logic and metaphysics set the standard in medieval universities and he was to have significant influence on later Catholic thought. His teachings persisted into the Renaissance, then to be vilified by the free-thinking Humanists, with stubborn adherents to Duns deemed 'old barking curs'. They became known as *dunsmen*. Thence came the word, *Dunse*, and then *dunce*, to mean dull-witted, and so the name of the conical *dunce's cap*, placed on the head of the unsatisfactory pupil. The travails of fate now have him as the *Blessed* John Duns Scotus, with his beatification by Pope John Paul II in 1992.

---

[13]The appendage *Scotus* simply meant that he hailed from Scotland, and from the border village of Duns.

His bit-part in the story of irrational numbers lies in his use (possibly the first such) of incommensurability in arguing against a sustained belief from classical antiquity that the universe behaved in a cyclical manner. According to Cicero[14]:

> On the diverse motions of the planets the mathematicians have based what they call the Great Year, which is completed when the sun, moon and five planets having all finished their courses have returned to the same positions relative to one another. The length of this period is hotly debated but it must necessarily be a fixed and definite time.

A popular agreement was that the period of this Great (or Perfect) Year was 36,000 years,[15] computed as 360×100 with Ptolemy's *Almagest* giving the procession of the equinoxes as 1 degree in 100 years. Whatever the wait, at its most extreme the belief had the universe and most particularly life on earth returning to its 'original' state with all repeated again – and again. This ultimate version of reincarnation was tidy and convenient to some (for example, the Stoics and their later adherents) as it was inconvenient to others (for example, the Christians) and the intervention of the intellectual Franciscan theologian Scotus is hardly surprising, although its nature was novel:

> This opinion[16] can also be disproved with respect to its cause, for if it could be proven that some celestial motion was incommensurable with another... then, I say, it follows that all the motions will never return to the same place.

He had replaced philosophical arguments with a mathematical criterion: the cyclic nature of the celestial objects required their motions to be *commensurable* with one another; a proof that this was not so would destroy the possibility of exact repetition and so the Perfect Year. And he gave a demonstrative, if theoretical, example. Suppose that two bodies move back and forth at the same speed, one along the side and the other the diagonal of a square (with implied instantaneous change of direction); if they each start at the same corner at the same time, then their simultaneous return to that corner would require the existence of positive

---

[14]Cicero, *On the Nature of the Gods* (Rackham translation).

[15]So, for other reasons, was 48,000 years.

[16]The cyclical nature of the universe.

integers m and n so that $n\sqrt{2} = m \times 1$, with the irrationality of $\sqrt{2}$ rendering this impossible. He conceded, though, that taking this argument to the motions of the celestial bodies would require a *great discussion*.

That great discussion was undertaken by many and none more respected than a man who was confessor to Edward III and (briefly) Archbishop of Canterbury. Chaucer, in the *Nun's Priest's Tale*, placed him in weighty intellectual company with the lines[17]:

> But I ne kan nat bulte it to the bren
> As kan the hooly doctour Augustyn,
> Or Boece, or the Bisshop Bradwardyn

Thomas Bradwardine (ca. 1290–1349 C.E.), the *Profound Doctor*, was also a noted scientist and among his varied publications there is the *Treatise on the Proportions of Velocities in Motion* of 1328, where he took up Aristotle's ideas from *Physics*, amended them and extended them. Aristotle believed that the ratio of force to resistance on an object is proportional to the velocity it thereby attains, a view which Bradwardine contradicted and replaced with (an equally false but neat) alternative that when that proportion is squared, cubed, etc., the associated velocity is doubled, tripled, etc.; that is, a geometric increase in the ratio of force to resistance results in an arithmetic increase in velocity.

In symbols, if the force F_1 is subject to a resistance R_1 and gives rise to a velocity v_1, we will write the ratio velocity association as $F_1/R_1 \rightarrow v_1$. If this ratio is squared to yield $F_2/R_2 = (F_1/R_1)^2$ we have according to Bradwardine $F_2/R_2 \rightarrow v_2$, where $v_2 = 2v_1$, which can be written $F_2/R_2 = (F_1/R_1)^{v_2/v_1}$.

The ideas were taken up and properties of the ratio investigated by a fellow *Oxford Calculator* of Merton College, Richard Swineshead (in his *Book of Calculations* of about 1350), but it with the thoughts of another that we are interested and so introduce a third contemporary luminary: Nicole Oresme (1323–1382 C.E.). Scholar, translator, philosopher, physicist, musicologist, economist, psychologist, author, regal councillor, mathematician, theologian and bishop, he was much taken with the

---

[17] But I cannot separate the valid and invalid arguments
As can the holy doctor Augustine
Or Boethius, or the Bishop Bradwardyn.

concept of incommensurability. And he was an equally determined opponent of astrology, which intrinsically depends on the precise positioning of the celestial objects.

Consistent with Scotus, his approach was to argue that, if the velocities of two celestial objects were incommensurable with each other, then so would be the distance they would travel in equal time periods, and this would render them returning to some initial configuration at some future time impossible. In terms of the Bradwardine equation for the ratios of the (unknown) forces and resistances he hoped for the exponent that is the velocity fraction to be an irrational number.

Of course, there was no hope of anything approaching a rigorous demonstration of this and he resorted to an almost touching leap of faith, extrapolating from a convenient finite arithmetic case to the vast complexities of celestial motion. His arguments first appeared in his work *De Proportionibus Proportionum* (*The Ratio of Ratios*) of the 1350s, where he took a notional 100 rational ratios from $\frac{1}{1}$ to $\frac{100}{1}$ for the F/R and counted the number of pairs of them which are related by a rational power. These start with the sequence

$$\frac{2}{1}, \left(\frac{2}{1}\right)^2, \left(\frac{2}{1}\right)^3, \left(\frac{2}{1}\right)^4, \left(\frac{2}{1}\right)^5, \left(\frac{2}{1}\right)^6$$

with $2^7 > 100$. That is, for example,

$$\left(\frac{2}{1}\right)^4 = \left(\left(\frac{2}{1}\right)^5\right)^{4/5} \quad \text{or} \quad \frac{16}{1} = \left(\frac{32}{1}\right)^{4/5}$$

and there are, of course, $\binom{6}{2} = 15$ of these.

The similar sequence for 3 yields

$$\frac{3}{1}, \left(\frac{3}{1}\right)^2, \left(\frac{3}{1}\right)^3, \left(\frac{3}{1}\right)^4$$

with $3^5 > 100$, and there are $\binom{4}{2} = 6$ of these.

One at a time, the remaining four possibilities arise from the set of pairings:

$$\frac{5}{1}, \left(\frac{5}{1}\right)^2; \frac{6}{1}, \left(\frac{6}{1}\right)^2; \frac{7}{1}, \left(\frac{7}{1}\right)^2; \frac{10}{1}, \left(\frac{10}{1}\right)^2.$$

In all, $15 + 6 + 4 \times 1 = 25$ are related by the expression $a/b = (c/d)^{m/n}$ for rational m/n. Yet, he argued, there are 4950 (= $\binom{100}{2}$) possible pairings and so $4950 - 25 = 4925$ for which the exponent must be irrational, resulting in the ratio

Rational pairings : Irrational pairings $= 4925 : 25 = 197 : 1$.

From this he inferred that, if greater numbers of ratios were taken the proportion of rational pairings would decrease, substantiated by the argument:

> However many numbers are taken in a series, the number of perfect or cube numbers is much less than other numbers and as more numbers are taken in the series the greater is the ratio of non-cube to cube numbers or non-perfect to perfect numbers. Thus, if there were some number and such information as to what it is or how great it is, and whether it was large or small, were wholly unknown... it would be likely that such an unknown number would not be a cube number or a perfect number.

Yet the proportion of force to resistance of the celestial bodies is unknown and surely there are many such governing their motion and so, most likely, the ratio of the velocities of any pair of them is irrational: the Perfect Year became eternity and the astrologers charlatans. From the same document:

> Moreover, incommensurability can be found in every kind of continuous thing, and in all instances in which continuity is imaginable, either extensively or intensively. For a magnitude can be incommensurable to a magnitude, an angle to an angle, a motion to a motion, a speed to a speed, a time to a time, a ratio to a ratio, a degree to a degree and a voice to a voice, and so on for any similar things.

This desperate argument assuredly failed to destroy the idea of a Perfect Year, but from it we might consider it as a first statement that the irrationals outnumber the rationals. Now we move forward to the sixteenth century, to find abundant evidence of the confident manipulation of surd irrationals arising from trigonometric values.

A Great Challenge

The challenge to which we refer concerns the polynomial

$$P(x) = 45x - 3{,}795x^3 + 95{,}634x^5 - 1{,}138{,}500x^7$$
$$+ 7{,}811{,}375x^9 - 34{,}512{,}075x^{11} + 105{,}306{,}075x^{13}$$
$$- 232{,}676{,}280x^{15} + 384{,}942{,}375x^{17} - 488{,}494{,}125x^{19}$$
$$+ 483{,}841{,}800x^{21} - 378{,}658{,}800x^{23} + 236{,}030{,}652x^{25}$$
$$- 117{,}679{,}100x^{27} + 46{,}955{,}700x^{29} - 14{,}945{,}040x^{31}$$
$$+ 3{,}764{,}565x^{33} - 740{,}259x^{35} + 111{,}150x^{37}$$
$$- 12{,}300x^{39} + 945x^{41} - 45x^{41} + x^{45}.$$

And it was not 'simply' a matter of finding its roots, but the fourfold question:

1. If

$$P(x) = \sqrt{2 + \sqrt{2 + \sqrt{2 + \sqrt{2}}}} \quad \text{show that} \quad x = \sqrt{2 - \sqrt{2 + \sqrt{2 + \sqrt{3}}}}.$$

2. If

$$P(x) = \sqrt{2 - \sqrt{2 - \sqrt{2 + \sqrt{2 + \sqrt{2}}}}}$$

show that

$$x = \sqrt{2 - \sqrt{2 + \sqrt{2 + \sqrt{2 + \sqrt{3}}}}}.$$

3. If

$$P(x)$$
$$= \sqrt{2 + \sqrt{2}}$$
$$= \sqrt{3.41421356237309504880168872420969807856967187 5375}$$

show that

$$x = \sqrt{2 - \sqrt{2 + \sqrt{\frac{3}{16}} + \sqrt{\frac{15}{16}} + \sqrt{\frac{5}{8}} - \sqrt{\frac{5}{64}}}}$$
$$= \sqrt{0.00274093049085225243101588312112683881 80}.$$

With the fourth part, to find x if

$$P(x) = \sqrt{\frac{7}{4} - \sqrt{\frac{5}{16}} - \sqrt{\frac{15}{8} - \sqrt{\frac{45}{64}}}}.$$

The poser was one Adriaan van Roomen[18] (1561–1615), with the challenge appearing at the end of the Preface of *Ideae Mathematicae*, his 1593 compilation of the most important living mathematicians of the time. Van Rooman, a significant Belgian mathematician (in particular he had used geometric methods to calculate π to 16 decimal places), was then professor of mathematics at the University of Louvain, which is located in the Flemish region of Belgium. It would seem that the challenge, issued to 'all the mathematicians throughout the whole world', was not expected to be answered by a Frenchman, since not a single French name had appeared in his list; but this was to omit Francois Vièta (1540–1603). It is true that Vièta was a professional lawyer and an amateur mathematician, but one capable enough to have decrypted an intercept of 1590 belonging to Philip II of Spain on behalf of his own king, Henry IV. His other mathematical contributions were variously significant but we see in his infinite product representation of π of 1593 (the earliest known)

$$\frac{2}{\pi} = \frac{\sqrt{2}}{2} \cdot \frac{\sqrt{2 + \sqrt{2}}}{2} \cdot \frac{\sqrt{2 + \sqrt{2 + \sqrt{2}}}}{2} \cdots$$

that he was a man at ease with complex surd expressions. In an audience, the Dutch ambassador had passed to Henry IV the unhappy news of the absence of French names in van Roomen's list. The king summoned his codebreaker.

Vièta had reached the infinite product expression for π using geometric arguments combined with the trigonometric identity $\cos 2\alpha = 2\cos^2\alpha - 1$, and it was this that he was to put to use once more, enabling him not only to find the single solution suggested but all positive ones.

In modern notation we will consider the solution of the first problem, which typifies his solution to them all – and begin with

---

[18]Often known by the Latin form of his name, *Adrianus Romanus*.

the right-hand side

$$\sqrt{2 + \sqrt{2 + \sqrt{2 + \sqrt{2}}}}.$$

The 'double angle formula' can be rewritten

$$2 \cos \alpha = \sqrt{2 + 2 \cos 2\alpha}$$

and from this we can set up a cascade of angles as follows:

$$2 \sin \frac{15\pi}{32} = 2 \sin \left(\frac{\pi}{2} - \frac{\pi}{32} \right)$$

$$= 2 \cos \frac{\pi}{32} = \sqrt{2 + 2 \cos \frac{\pi}{16}}$$

$$= \sqrt{2 + \sqrt{2 + 2 \cos \frac{\pi}{8}}}$$

$$= \sqrt{2 + \sqrt{2 + \sqrt{2 + 2 \cos \frac{\pi}{4}}}}$$

$$= \sqrt{2 + \sqrt{2 + \sqrt{2 + \sqrt{2}}}}.$$

With the right-hand side of the equation identified in terms of a trigonometric function, we can follow Vièta in the far greater task of doing the same with the left-hand side, first using another standard trigonometric identity: $\sin 3\alpha = 3 \sin \alpha - 4\sin^3 \alpha$.
From this we conclude that, if

- $\alpha = 15\theta$, $\sin 45\theta = 3 \sin 15\theta - 4\sin^3 15\theta$;
- $\alpha = 5\theta$, $\sin 15\theta = 3 \sin 5\theta - 4\sin^3 5\theta$.

A final trigonometric identity of use is

$$\sin^5 \theta = \tfrac{5}{8} \sin \theta - \tfrac{5}{16} \sin 3\theta + \tfrac{1}{16} \sin 5\theta,$$

which, combined with the one above, makes

$$\sin 5\theta = 16\sin^5 \theta - 10 \sin \theta + 5 \sin 3\theta$$

$$= 16\sin^5 \theta - 10 \sin \theta + 5(3 \sin \theta - 4\sin^3 \theta)$$

and so

$$\sin 5\theta = 16\sin^5 \theta - 20\sin^3 \theta + 5 \sin \theta.$$

Therefore,

$$\sin 15\theta = 3 \sin 5\theta - 4\sin^3 5\theta$$
$$= 3[16\sin^5\theta - 20\sin^3\theta + 5 \sin\theta]$$
$$- 4[16\sin^5\theta - 20\sin^3\theta + 5 \sin\theta]^3,$$

which makes

$$\sin 45\theta = 3 \sin 15\theta - 4\sin^3 15\theta$$
$$= 3\{3[16 \sin^5\theta - 20 \sin^3\theta + 5 \sin\theta]$$
$$- 4[16 \sin^5\theta - 20 \sin^3\theta + 5 \sin\theta]^3\}$$
$$- 4\{3[16 \sin^5\theta - 20 \sin^3\theta + 5 \sin\theta]$$
$$- 4[16 \sin^5\theta - 20 \sin^3\theta + 5 \sin\theta]^3\}^3,$$

which is a mess: yet a mess which, as a function of $\sin\theta$, has order $5 \times 3 \times 3 = 45$.

   If we make the substitution $x = 2 \sin\theta$ into the expression and (get a computer to) simplify, we find that the result is precisely $\frac{1}{2}P(x)$. That is, we have

$$P(x) = 2 \sin 45\theta, \quad \text{where } x = 2 \sin\theta.$$

Finally, then, to solve the original algebraic equation is to solve the trigonometric equation

$$2 \sin 45\theta = 2 \sin\left(\frac{15\pi}{32}\right)$$

to get

$$45\theta = \frac{15\pi}{32} + 2k\pi$$

and so

$$\theta = \frac{15\pi}{32 \times 45} + \frac{2k\pi}{45} = \frac{\pi}{3 \times 2^5} + \frac{2k\pi}{45}, \quad k = 0 \pm 1, \pm 2, \ldots,$$

and return to the x values of

$$x = 2 \sin\left(\frac{\pi}{3 \times 2^5} + \frac{2k\pi}{45}\right), \quad k = 0 \pm 1, \pm 2, \ldots$$

These evaluate to the approximations

{0.0654382, 0.342998, 0.613883, 0.872818, 1.11477, 1.33502,
1.52928, 1.69378, 1.82531, 1.92132, 1.97992, 1.99999,
1.98114, 1.92372, 1.82886, 1.6984, 1.53489, 1.3415,
1.122, 0.880662, 0.622182, 0.351593, 0.0741595,
$-$ 0.204717, $-$0.479609, $-$0.745166, $-$0.996219,
$-$ 1.22788, $-$1.43565, $-$1.61547, $-$1.76384,
$-$ 1.87789, $-$1.95538, $-$1.99482, $-$1.99543,
$-$ 1.9572, $-$1.88087, $-$1.76794, $-$1.62059, $-$1.44171,
$-$ 1.23476, $-$1.00378, $-$0.753257, $-$0.488076, $-$0.213396}.

Vièta had found the first two solutions immediately, the remainder of the 23 the following day, but the final 22 were ever to elude him: dealing with irrational quantities was one thing, dealing with negative ones quite another. The interested reader may wish to follow Vièta's path with the remaining three parts of the challenge, the right-hand sides of which may be checked to be

$$2 \sin \left(\frac{15\pi}{64}\right), \quad 2 \sin \left(\frac{3\pi}{8}\right) \quad \text{and} \quad 2 \sin \left(\frac{\pi}{15}\right)$$

respectively and then reference his own original account, which he had published in 1595 as *Ad Problema. Quod omnibus Mathematicis totius orbis construendum proposuit Adrianus Romanus*, and which is readily available.

Before we leave this impressive control over trigonometric surd manipulation we shall put the problem in context with the non-trivial observation that

$$P(x) = \sum_{i=0}^{22 = \lfloor \frac{45}{2} \rfloor} (-1)^i \frac{45}{45 - i} \binom{45 - i}{i} x^{45-2i},$$

which allows the generalization

$$P_n(x) = \sum_{i=0}^{\lfloor n/2 \rfloor} (-1)^i \frac{n}{n - i} \binom{n - i}{i} x^{n-2i}.$$

It transpires that, for n odd, we have $P_n(2 \sin \theta) = 2 \sin(n\theta)$. The periodicity of $\sin \theta$ ensures that $\sin(n\theta) = \alpha$ has precisely n roots

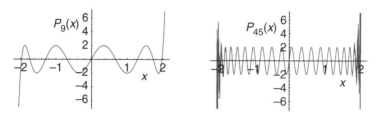

Figure 2.2.

and so the parameterization of $x = 2 \sin \theta$ accounts for them all. Figure 2.2 displays the behaviour of $P_9(x)$ and $P_{45}(x)$.

A Cloud of Infinity

We find in the work of the Italian Franciscan friar Luca Pacioli a study of the Golden Ratio, entitled *De Divina Proportione*. Famous though the book is, it relied heavily on the *Liber Abaci* and is no more than a compilation of known facts; it added nothing to the understanding of irrationality. The reader will glean some idea of its limited significance in this story from the following excerpt:

> [J]ust like God cannot be properly defined, nor can be under-stood through words, likewise this proportion of ours cannot ever be designated through intelligible numbers, nor can it be expressed through any rational quantity, but always remains occult and secret, and is called irrational by the mathematicians.

More insightful is the admittedly confused view of the greatest German algebraist of the sixteenth century, the Augustinian monk Michael Stifel (1487–1567). In his *Arithmetica Integra* of 1544 he stated:

> It is rightly disputed whether irrational numbers are true numbers or false. Since, in studying geometrical figures, where rational numbers fail us irrational numbers take their place and prove exactly those things that rational numbers could not prove... we are moved and compelled to admit that they truly are numbers, compelled that is by the results that follow from their use – results that we perceive to be real, certain and constant. On the other hand, other considerations compel us to deny that irrational numbers are numbers at all. To wit, when we seek to subject them to numeration we find that they flee

away perpetually, so that not one of them can be apprehended precisely in itself... Now that cannot be called a true number which is of such a nature that it lacks precision... Therefore, just as an infinite number is not a number, so an irrational number is not a true number, but lies hidden in a kind of cloud of infinity.

As for π:

Therefore the mathematical circle is rightly described as the polygon of infinitely many sides. And thus the circumference of the mathematical circle receives no number, neither rational nor irrational[19].

Yet, this most influential work is distinguished for many reasons, not least of which is the degree to which the algebraic method is used, together with its associated modern algebraic notation such as $+$, $-$ and $\sqrt{\ }$; moreover, we find in it his development of exponents and his own version of logarithms. His further work embraced the irrationals of Book X of *The Elements* and he considered expressing irrationals in decimal notation. We will leave him, though, with his arithmetically supported mysticism. A committed follower of his friend Martin Luther, he used his own form of numerology to prove that the name of Pope Leo X concealed the *number of the beast*; 666 in the Book of Revelation. His method was to write the pope's name in convenient form, extract from it the Roman numerals and perform convenient manipulations on them, as follows:

$$\text{Leo X} \to \text{LEO DECIMVS} \to \text{LeoDeCIMVs}$$
$$\to \text{LDCIMV} \to \text{MDCLVI} \to 1656.$$

Then, remove the M, the first letter of *Mysterium*, and add in the X of Leo X to get $656 + 10 = 666$[20]: now, that *is* irrational.

---

[19]Meaning, *algebraic*.

[20]The ire of the Catholic reader may be assuaged by the efforts of the contemporary Catholic theologian, Peter Bongus, who used different letter assignments and a similar method to prove that Luther was the antichrist. Frustrated Microsoft users might be interested in the efforts of *Harper's* magazine, which identified Bill Gates as the antichrist using ASCII encoding.

It is, perhaps, not surprising that after his numerological methods[21] predicted that the world would end at 8 a.m. on 19 October 1533 (the catastrophic failure causing him to be held in protective custody) no further record is found of his use of them.

But finally we should acknowledge the influential insight of the Dutch mathematician, engineer and accountant, Simon Stevin (1548–1620). It was he who introduced the decimal numbering system (of the Arabs and Hindus, as well as the Chinese) to Europe through his 1585 book, written in Dutch, *La Thiende*. Immediate translations into French (*La Disme*) and English (*Disme, The Arts of Tenths or Decimal Arithmetike*) followed, bringing the ideas to the greater European world and the decimal place system to the heart of mathematics. Restricted though it was (only finite decimals were allowed), it was embraced if not universally then significantly, most particularly by John Napier with his tables of logarithms. The year was clearly a busy one for Stevin for (apart from much else) he also published (in French) *L'arithmétique*, really a compendium volume of a collection of short books, each related in something of a loose way, but all centred on the essential idea of the arithmetic of integers, fractions and irrationals, as well as something of the algebra of polynomials and theory of equations. For him 1 was just another number but 0 was not recognized, although negative numbers were tolerated (with considerable suspicion) but complex numbers were entirely unacceptable. Further, he made the point that the Euclidean idea of number was too restrictive and that number is a continuous concept:

> ...as continuous water corresponds to a continuous humidity, so does a continuous magnitude correspond to a continuous number...

and, less prosaically:

> There are no absurd, irrational, irregular, inexplicable or surd numbers... It is true that $\sqrt{8}$ is incommensurable with an arithmetic number, but this does not mean it is absurd... if $\sqrt{8}$ and an arithmetical number are incommensurable then it is as much (or as little) the fault of $\sqrt{8}$ as it is the arithmetical number.

---

[21]In this case, identifying letters with triangular numbers, so inventing the *trigonal* alphabet.

Progress, then, and progress was becoming a pressing matter as mathematics developed. We will consider one particularly significant aspect of it.

The Punctured Plane

The Frenchmen Pierre de Fermat (1601–65) and René Descartes (1596–1650) had little more in common than their nationality. Fermat the diffident jurist and amateur mathematician, Descartes the philosopher and egotistical sceptic; the one most commonly remembered for his *Last Theorem*, the other for his statement *Cogito ergo sum*[22], they were contemporaries but only in time, never met but each independently invented analytic geometry.[23] Their approaches contrasted as their lives contrasted with Fermat commonly writing down an equation in x and y and then investigating its associated curve, Descartes commonly picking a known curve and finding its x, y equation. Yet, reading their (translated) work makes clear that the idea of a pair of perpendicular axes was not clearly formed and that, such reference as there is, was made to a pair of oblique directions and then in something of an oblique manner: the view of George F. Simmons has merit[24]:

> Superficially, Descartes' essay looks as if it might be analytic geometry, but isn't; while Fermat's doesn't look it, but is.

Never mind, though, the historiological comparisons: with the advent of analytic geometry came the necessity to solve equations, which were the equivalent of their associated geometric problems, and these would often have irrational roots; hardly a new problem, but one which now assumed far greater practical moment. Take, for example, the fundamental problem of finding the distance between two points in the plane. With the prevalence of quadratic irrationalities, even though the points themselves might have rational coordinates, the distance between them $\sqrt{(x_1 - x_2)^2 + (y_1 - y_2)^2}$ would likely be irrational.

Now suppose that we tactically avoid irrational numbers altogether and so consider only rational distances and rational

---

[22]"I think therefore I am."

[23]The amalgamation of geometry with arithmetic, algebra and analysis through the use of a coordinate system.

[24]George F. Simmons, 1992, *Calculus Gems* (McGraw-Hill).

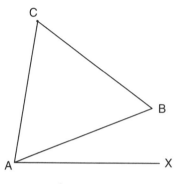

Figure 2.3.

points[25] in the plane. Now suppose that we can choose three such points to form the sides of a triangle with rational sides, then the cosine rule with standard notation has it that

$$\cos A = \frac{b^2 + c^2 - a^2}{2bc}$$

is rational. But we will see in chapter 4 that, if A is rational and measured in degrees, the only rational values of $\cos A$ are $0, \pm\frac{1}{2}, \pm 1$: the only acceptable values for A are then $60°$ and $90°$ and, since the same must be true for all three angles, we are left with the common angle $A = 60°$ and the triangle being equilateral. In short, the only possible triangles with rational sides must be equilateral. But the vertices of our triangle are rational points, which means that in figure 2.3 the gradients of sides AB and AC are rational numbers m_1 and m_2 respectively and using a standard trigonometric identity:

$$\tan \angle CAB = \tan(\angle CAX - \angle BAX)$$
$$= \frac{\tan \angle CAX - \tan \angle BAX}{1 + \tan \angle CAX \tan \angle BAX} = \frac{m_2 - m_1}{1 + m_1 m_2}.$$

This means that $\tan \angle CAB$ is a rational number and again we will see in chapter 4 that this makes $\angle CAB = 45°$.

We are inexorably trapped in a logical and arithmetical maelstrom: triangles cannot exist.

---

[25] A rational point in the plane is one both coordinates of which are rational numbers.

Now move to the unit circle $x^2 + y^2 = 1$. With its standard algebraic parameterization,

$$x = \frac{1 - t^2}{1 + t^2}, \qquad y = \frac{2t}{1 + t^2},$$

we know that it contains an infinite number of rational points, with the parameterization yielding all of them as the parameter t varies over all rational numbers. Yet there are an infinite number of points on the circle at least one coordinate of which is irrational. Does this matter? Attempting to find the intersection of $x^2 + y^2 = 1$ with the line $y = x$ quickly reveals that irrational numbers cannot be avoided.

And what of the circle $x^2 + y^2 = 5$? Again it has an infinite number of points at least one coordinate of which is an irrational number but again it boasts an infinite number of rational points too, which the reader can check are generated by the parameterization[26]

$$x = \frac{t^2 - 4t - 1}{t^2 + 1}, \qquad y = -2\frac{t^2 + t - 1}{t^2 + 1}$$

for rational t.

And for $x^2 + y^2 = 3$? Are we merely changing the radius to change the parameterization? The answer is no, since, without the existence of irrational numbers, this circle would not appear at all in Descartes's plane: there is not a single rational point on it; there can be no rational parameterization. And we can prove this.

Assume that rational point $(p/q, r/s)$ lies on the circle and so $p^2/q^2 + r^2/s^2 = 3$, resulting in $p^2 s^2 + q^2 r^2 = 3q^2 s^2$ and so $(ps)^2 + (qr)^2 = 3(qs)^2$ and therefore the existence of positive integers so that $a^2 + b^2 = 3c^2$. Since the sum of two even numbers and the sum of two odd numbers are each even, it must be that one of a^2 and b^2 is even and the other odd; so write $a^2 = 2m$ and $b^2 = 2n + 1$, then $a^2 + b^2 = 4(m^2 + n^2 + n) + 1 = 4N + 1 = 3c^2$. Now let c be any of the four possibilities $c = 4M, 4M + 1, 4M + 2, 4M + 3$, which means that c^2 is of the form $4N, 4N + 1, 4N, 4N + 1$ respectively, and so $3c^2$ must be of the form $4N, 4N + 3, 4N, 4N + 3$

---

[26]The interested reader should turn to Appendix B on page 278 for the derivation of this.

respectively, with the required $4N + 1$ missing: a contradiction. The Fermat *Sum of Two Squares* result

> A positive integer n can be written as the sum of two squares if and only if each of its prime factors of the form $4k + 3$ occurs to an even power.

assures the existence of plenty more problematic circles. And, of course, circles are simply special cases of problems that must exist in the punctured plane of rational points: the disappearance of the cubic curves $x^3 + y^3 = 1$ and $x^3 + y^3 = 3$ is assured, as the characterization of integers that can be written as the sum of two cubes translates to its geometric equivalent. And we could continue indefinitely with varied examples of curves either partially or completely disappearing, but will end with just one more: e^x would appear as the single rational point $(0, 1)$.

In short it was inescapable that, little understood though they were, with the advent of Cartesian coordinates and so the algebraicization of geometry, irrational numbers were at least implicitly and often explicitly indispensable; without them life in the Cartesian plane is unsustainable. And differential and integral calculus were soon to move from infancy to young maturity; let us now focus on part of that infancy.

Descartes's approach to analytic geometry appeared in his famous 1637 publication *Discours de la méthode pour bien conduire sa raison, et chercher la vérité dans les sciences,*[27] usually referred to as the *Discourse on Method*, with the title disclosing its purpose. To be exact, it appears in *La Géométrie*, the third of three appendices to the main work, the purpose of which were to exhibit fruits of the application of his 'method'. In fact, *La Géométrie* is itself divided into three 'books', the second of which was a precursor to the differential calculus since it contains his *method of normals*, which is his approach to finding the gradient of a curve at any given point. A curve was to be the trace of a pair of what we now call parametric equations, but not just any pair since

> the spiral, the quadratix and similar curves

were to be neglected for being *mechanical* curves which belong not to the world of geometry, as do the *geometric* curves, but

---

[27] *Discourse on the Method of Rightly Conducting One's Reason and of Seeking Truth in the Sciences.*

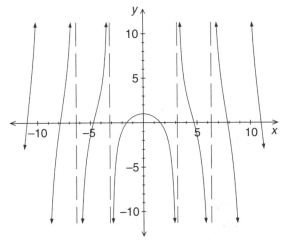

Figure 2.4.

to the physical world – which is something quite different. It is, perhaps, particularly unfair to so condemn the quadratrix, since it was this curve (according to Proclus) that Hippias of Elis (ca. 460 B.C.E.) used in his investigations of the ancient problems of trisecting the angle and squaring the circle (which we will mention again in chapter 7). It may have been the first curve to be identified which does not arise from straight lines and circles and the first after them to be given a name; it is shown in figure 2.4.

Unfortunately, Descartes' distinction between mechanical and geometric curves is a little blurred, with his defining criterion for a mechanical curve being

> since they must be conceived of as being described by two separate movements whose relation does not admit of exact determination.

The Cartesian equation of the quadratrix can be written $y = x \cot x$. Its simplest parametric alternative of $y = t \cot t, x = t$ has an essential reliance on trigonometric functions, which were to become known as examples of *transcendental functions*; that is, functions which transcend algebraic methods. And Descartes needed algebraic methods for his purposes. Of course, the unit circle $x^2 + y^2 = 1$ has a standard trigonometric parameterization $x = \cos t, y = \sin t$ but we have already noted that it also admits an alternative algebraic form; it is the difference

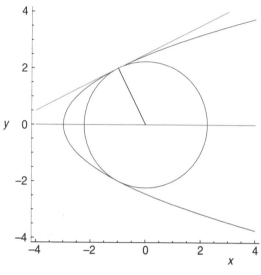

Figure 2.5.

between the essential reliance on non-algebraic functions or not that distinguishes between his two types of curve.

His principle for finding the gradient of a curve at a point is initially appealing: fit a circle (the centre of which he supposed to lie on the x-axis) to the curve so that it touches the curve at the given point; it is easy, then, to compute the gradient of the radial line from the centre of the circle to the point (the normal) and so the gradient of the line perpendicular to it (the tangent to the circle and to the curve at the point), as shown in figure 2.5. For example, let us use his method to find the gradient at the point $(-1, 2)$ of the tangent to the parabola given in parametric form as $x = \frac{1}{2}t^2 - 3$, $y = t$.

Its Cartesian form is evidently $y^2 = 2x + 6$ and we must find the centre and radius of the circle which suits purpose and so begin with its equation $(x - a)^2 + y^2 = r^2$. Throughout, we will refer to figure 2.5 and see that the radius of the circle must be the distance between $(a, 0)$ and $(-1, 2)$ and this means that we have

$$(x - a)^2 + y^2 = (a + 1)^2 + 2^2.$$

At the intersection we can substitute for y to get

$$(x - a)^2 + 2x + 6 = (a + 1)^2 + 2^2,$$

which simplifies to the quadratic equation

$$x^2 + (2 - 2a)x + (1 - 2a) = 0,$$

which, for our purpose, must have equal roots, and so

$$(2 - 2a)^2 = 4(1 - 2a)$$

resulting in $a = 0$ as the only possible value. The required circle is, then, one we have looked at before: $x^2 + y^2 = 5$. The gradient of the normal is $\frac{2}{-1} = -2$, making the gradient of the tangent to the parabola at the given point $\frac{-1}{-2} = \frac{1}{2}$.

Surely his approach works, at least in theory, and after all it is kin to the modern approach of defining the curvature of any given curve, but in practical terms it was limited – and much more limited than the alternative suggested by Fermat. Our example has been chosen carefully so that matters work out, as were those provided by Descartes, all of which conceal the inconvenient fact that the calculations involved in any particular case can easily be formidable (not least, involving irrational numbers), or to frame matters in the words of a contemporary, they could be

> a labyrinth from which it is extraordinarily difficult
> to emerge.

Fermat's own method, although hardly rigorous or clearly explained, can be reduced to a now very familiar symbolic form from elementary calculus:

$$f'(a) \approx \frac{f(a + h) - f(a)}{h},$$

where h is taken to be 'small'.

We are now as clearly aware of the worth of this approach as Descartes was ignorant of it and the innocent publication by Fermat of his alternative immediately led to one of those unfortunate public disputes of priority and probity, with Descartes's considerable ego much affronted. It is amusing that when Descartes insisted that Gérard Desargues, whose own work was hardly known for its lucidity, referee the dispute he formed the view that

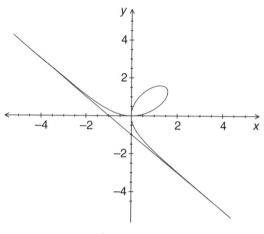

Figure 2.6.

Messieur Descartes is right and Messieur Fermat is not wrong.

A charming diplomatic judgement which failed to satisfy the enraged Descartes and which helped to bring to mathematics one of its famous algebraic curves, now known as the *Folium of Descartes*, and which has parametric equations

$$x = \frac{3at}{1+t^3}, \qquad y = \frac{3at^2}{1+t^3}$$

and Cartesian equation

$$x^3 + y^3 = 3axy$$

for any constant a. A typical Cartesian view of it is shown in figure 2.6 and Descartes challenged Fermat to find its gradient at any point, having failed to do so himself. Fermat's near-instant solution brought from Descartes a final grudging acceptance of his method. Acceptance there may have been but with it the question of a limit, which had simmered since ancient times, demanded attention – and limiting processes demand continuity and continuity demands the existence of irrational numbers. For differential calculus ever to be set on a firm foundation the number line and plane required microscopic attention, which they were not to get for two more centuries.

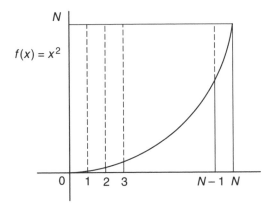

Figure 2.7.

Catching Proteus

To date, the longest occupation of the Savilian Chair of Geometry at Oxford is that of John Wallis (1616–1703), with a tenure
of some 54 years lasting from 1649 to his death. We have just
mentioned one dispute and, although he was the most influential English mathematician before Newton and inspiration to him,
it is easy to remember Wallis solely for another: the twenty-year
feud with the philosopher Thomas Hobbes (over their divergent
views on the problem of 'squaring the circle'). Yet we will single
out his approach to the other half of calculus, *quadrature*, or what
we would now call integration, as a further significant episode in
the problems associated with the continued frail understanding
of irrational numbers. Notwithstanding Hobbes's view of 'your
scurvy book *Arithmetica infinitorum*', it is this 1656 publication
of Wallis that bequeaths to us his most significant mathematical
thoughts.

By the seventeenth century the problem of quadrature, that is,
finding the area of a plane figure or the volume of a solid one,
was subject to two methods of approach: the method of exhaustion, handed down by the ancients (notably Eudoxus, Archimedes
and Pappas), and the use of infinitesimals, again studied by the
ancients but largely abandoned following the inexplicable para-
doxes of Zeno. Kepler had resurrected the use of infinitesimals
in his *Stereometria* of 1615, to be followed by Cavalieri, Roberval,
Pascal, Fermat and Torricelli, who between them had established

what we would write as

$$\int_0^1 x^p \, dx = \frac{1}{p+1}$$

for p a positive integer. In *Arithmetica infinitorum*,[28] Wallis adopted a variant of the Cavalieri Principle,[29] which, in two dimensions, can be described in modern terms in the following way.

Given a positive-valued continuous function $f(x)$ defined on the positive real axis and having an absolute maximum value of M_N on the interval $[0, N]$:

$$\int_0^1 f(x) \, dx = \lim_{N \to \infty} \frac{\sum_{r=0}^{N} f(r)}{\sum_{r=0}^{N} M_N} = \lim_{N \to \infty} \frac{\sum_{r=0}^{N} f(r)}{M_N(N+1)}.$$

In words, add up the lengths of the vertical strips from the x-axis to the curve and then divide this by the same order approximation of the area of the enclosing rectangle. We can look at a simple example, taking $f(x) = x^2$, as shown in figure 2.7:

$$\int_0^1 x^2 \, dx = \lim_{N \to \infty} \frac{\sum_{r=0}^{N} r^2}{N^2(N+1)} = \lim_{N \to \infty} \frac{\frac{1}{6}N(N+1)(2N+1)}{N^2(N+1)}$$

$$= \lim_{N \to \infty} \frac{1}{6}\left(\frac{2N+1}{N}\right) = \frac{1}{3}.$$

Most particularly, a great problem that had been considered for many years was the quadrature of the circle, the resolution of which would establish the nature of what we would call π; irrationals were still synonymous with surds and it was a matter of considerable moment to establish whether π was irrational or not and if so, how (if at all) it can be expressed in surd form: referring to figure 2.8, the ratio of the area of a circle to its squared radius can be computed (as $\pi/4$). With the circle's equation $x^2 + y^2 = R^2$ he took the special case of $R = 6$ to give the circle $x^2 + y^2 = 6^2$

---

[28] All quotations refer to *The Arithmetic of Infinitesimals: John Wallis 1656*, J. A. Stedall, *Sources & Studies in the History of Mathematics & the Physical Sciences* (Springer).

[29] Two solids are of equal volume if the areas of their cross-sections are correspondingly everywhere equal.

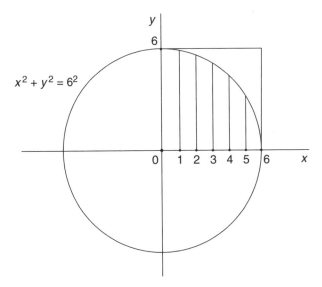

Figure 2.8.

and also $N = 6$ to compute the expression

$$\frac{\sqrt{36 - 0^2} + \sqrt{36 - 1^2} + \sqrt{36 - 2^2} + \sqrt{36 - 3^2} + \sqrt{36 - 4^2} + \sqrt{36 - 5^2} + \sqrt{36 - 6^2}}{6 + 6 + 6 + 6 + 6 + 6 + 6}$$

and simplified it to

$$\frac{6 + \sqrt{35} + 4\sqrt{2} + 3\sqrt{3} + 2\sqrt{5} + \sqrt{11} + 0}{42}.$$

And he repeated the calculation with $R = N = 10$ and simplified that resulting expression to

$$\frac{19 + \sqrt{19} + \sqrt{51} + 5\sqrt{3} + 2\sqrt{21} + \sqrt{911} + 4\sqrt{6} + 3\sqrt{11}}{105}.$$

The expression becomes ever more complicated with no discernable pattern in the surds and so no obvious route to simplification of them; that limit seemed literally far away and he commented:

> And indeed the more parts there are taken of the radius or diameter, so much less does the ratio of all the sines[30], to the

---

[30] Ordinates.

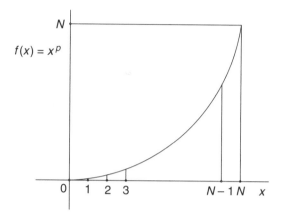

Figure 2.9.

greatest taken the same number of times, seem expressible. Therefore if the radius or diameter is taken in infinitely many parts (which it seems must be done for our purposes) the ratio of all the sines, to the radius taken the same number of times, that is, the quadrant or semicircle to the circumscribed square or parallelogram, seems wholly inexpressible, at least unless an expression of this kind is judged to be sufficient.

π's nature remained mysterious and its continued resilience to reveal its nature brought a rather touching paragraph from a man known for his curmudgeonly manner:

And thus having weighed this carefully, it nearly came about that I abandoned the investigation of the thing that, as it were, I so called for above. The one thing that gave hope was this. That is, that the same difficulty notwithstanding, in square roots, cube roots, biquadratic roots, etc. of numbers in arithmetic proportion the thing turned out not badly.

His retreat to simpler functions for his process brought about measured success but no little confusion.

We have mentioned that $\int_0^1 x^p \, dx = 1/(p+1)$ was already known for p a positive integer and, using $f(x) = x^p$ and what he called induction, he was satisfied with (referring to figure 2.9)

$$\frac{0^p + 1^p + 2^p + \cdots + N^p}{N^p + N^p + N^p + \cdots + N^p} \xrightarrow[N \to \infty]{} \frac{1}{p+1} = \int_0^1 x^p \, dx.$$

And he extrapolated from this using what he called *interpolation*, arguing that the result held true for a positive rational exponent and so $f(x) = x^{p/q}$ with the result

$$\frac{0^{p/q} + 1^{p/q} + 2^{p/q} + \cdots + N^{p/q}}{N^{p/q} + N^{p/q} + N^{p/q} + \cdots + N^{p/q}} \xrightarrow[N \to \infty]{} \frac{1}{p/q + 1} = \int_0^1 x^{p/q}\, dx.$$

And he extrapolated further still. Using what we shall call *blind optimism* he also asserted that the result holds for p irrational, stating with no hint of justification that

if $p = \sqrt{3}$ the emergent corresponding ratio is $\dfrac{1}{\sqrt{3}+1}$.

That is,

$$\frac{0^{\sqrt{3}} + 1^{\sqrt{3}} + 2^{\sqrt{3}} + \cdots + N^{\sqrt{3}}}{N^{\sqrt{3}} + N^{\sqrt{3}} + N^{\sqrt{3}} + \cdots + N^{\sqrt{3}}} \xrightarrow[N \to \infty]{} \frac{1}{\sqrt{3}+1} = \int_0^1 x^{\sqrt{3}}\, dx.$$

Returning to the seemingly safer ground of $f(x) = x^{p/q}$ and with $p/q = \tfrac{1}{2}$ he again took the case of $N = 6$ to achieve

$$\frac{\sqrt{0} + \sqrt{1} + \sqrt{2} + \sqrt{3} + \sqrt{4} + \sqrt{5} + \sqrt{6}}{7\sqrt{6}}$$

$$= \frac{3\sqrt{6} + 2\sqrt{3} + 3\sqrt{2} + \sqrt{30} + 6}{42}$$

and again made the point that increasing N *and similarly other series of this kind,* made the surd expression ever more complicated for any finite value of N:

> And this indeed seems to be confirmed still more strongly, since a finite series of this kind, to a series of the same number of terms equal to the greatest, has scarcely allowed any other expression of the ratio than by repetition of everything piece by piece; for rarely do two or more happen to be commensurable, that can be gathered into one sum.

Yet, in the limit the surds disappear.

> And indeed I was sometimes inclined to believe the thing to be quite impossible, that an infinite number of surd roots, incommensurable to each other, might be brought together in one sum that has an explicable ratio to some proposed rational quantity. But if the same series is supposed continued to infinity, it will eventually produce the ratio $\tfrac{2}{3}$ ... the infiniteness itself indeed (which seems amazing) destroying the irrationality.

That is,

$$\frac{\sqrt{0} + \sqrt{1} + \sqrt{2} + \cdots + \sqrt{N}}{\sqrt{N} + \sqrt{N} + \sqrt{N} + \cdots + \sqrt{N}}$$

$$= \frac{0^{1/2} + 1^{1/2} + 2^{1/2} + \cdots + N^{1/2}}{N^{1/2} + N^{1/2} + N^{1/2} + \cdots + N^{1/2}} \xrightarrow[N \to \infty]{} \frac{1}{\frac{1}{2} + 1}$$

$$= \frac{2}{3} = \int_0^1 \sqrt{x}\, dx.$$

The little understood concept of limiting processes combined with the scarcely better understood idea of irrationality posed a conundrum too involved for the age, in spite of the remarkable abilities of some of the mathematicians who lived in it.

And so Wallis, typically of his time, manipulated surds with abandon and progressed from the specific to the general through a mixture of his two techniques of induction and interpolation, which today are as properly defined and rigorous as they were otherwise for him. His tentative grasp on the nature of π was to cause him to coin a most appropriate metaphor.

> Although no small hope seemed to shine, what we have in hand is slippery, like Proteus, who in the same way, often escaped, and disappointed hope.

Proteus, the Old Man of the Sea of Greek legend, was all-knowing but refused to answer questions unless he was captured and held, a task made the more difficult by his ability to change into any shape imaginable.

The shape of π proved as elusive to Wallis, but we will leave him as he continues to set the foundations for the currently ten-year-old Newton (and eight-year-old Leibniz) to bring the differential and integral calculus to the world on a positive note. Whatever kind of irrational π may or may not be and unable though he was to write it in terms of surds, he managed in terms of positive integers with the Wallis Product,

$$\square = \frac{3 \times 3 \times 5 \times 5 \times 7 \times 7 \times \text{etc.}}{2 \times 4 \times 4 \times 6 \times 6 \times 8 \times 8 \times \text{etc.}} = \frac{9 \times 25 \times 49 \times \text{etc.}}{8 \times 24 \times 48 \times \text{etc.}}$$

where the symbol $\square$ was his way of writing what we write as $4/\pi$.

And with a final comment from him:

> The ratio of perimeter to diameter is neither a fraction nor a surd... there must be some other way of Notation invented[31] than either Negatives or Fractions, or (what are commonly called) Surd Roots, or the Roots of Ordinary Equations or even the Imaginary Roots of such Impossible Equations in the ordinary forms.

There are: transcendental numbers.

With Wallis we creep over the boundary into the eighteenth century. This was not a century given to mathematical hand-wringing, but one of expansive development: negative numbers may not have yet been on a comfortable foundation but their usefulness was manifest, so for that matter were complex numbers. As for irrationals, they were philosophically easier than negative numbers, and it was more urgent to establish whether a particular number was one rather than worry too much about what implications this might have; anyway, there was always the Axiom of Eudoxus upon which to fall back. Here is the period in which Leonhard Euler first proved e to be irrational and Johann Lambert the same for π. In 1794 Adrien-Marie Legendre published his *Eléments de Géométrie* in which he reorganized Euclid's *Elements* and in doing so replaced it as the standard geometric text of Europe. As with *The Elements*, the work did not confine itself to geometry alone and in it he gave a simpler proof that π is irrational as well as the first proof that π^2 is irrational; he also conjectured that π is not the root of any algebraic equation of finite degree with rational coefficients, putting in clearer terms the suspicion of Wallis.

We will consider these matters in the following chapter.

---

[31] Some other kind of number.

Two New Irrationals

It can be of no practical use to know that π is irrational, but if we can know, it surely would be intolerable not to know.

Ted Titchmarsh

We commented in the Introduction that "first proofs are often mirror-shy" and this chapter is devoted to two of them; neither can be placed "fairest of them all" and it is to great eighteenth-century (mathematical) kings that we look, not a wicked queen, as we gaze closely at them. In the end, it was not to be that π's mysterious nature was first to be understood, since this occurred fully thirty years after the irrationality of e was established – a number not born in antiquity but in the eighteenth century itself. We have seen Wallis struggle with the nature of π in the last chapter; he suspected that, not only was it irrational, but that it was a new type of irrational, not definable in terms of a finite numbers of roots; its irrationality we establish here but that second quality will have to wait until chapter 7, as it had to wait for two hundred years.

As we progress our investigation into the history of irrational numbers, the theory of *continued fractions* is of great moment. It is through these beautiful constructions that both π and e were first proved to be irrational and, in order to understand matters, we shall need to be secure about what they are and how to construct and manipulate them: their further role is discussed in chapter 8.

Elementary Continued Fractions

The study of continued fractions was by the eighteenth century reasonably old. The two near contemporaries from the Italian city of Bologna, Rafael Bombelli (1526-72) and Pietro Cataldi (1548–1626), had dealt with special cases but it is with the work of John Wallis that the topic began to reveal its potential. From his work others followed, with one particularly celebrated result that of the first president of the Royal Society, Lord Brouncker (1620-84), who used Wallis's infinite product from page 90 to show that

$$\frac{4}{\pi} = 1 + \cfrac{1^2}{2 + \cfrac{3^2}{2 + \cfrac{5^2}{2 + \cfrac{7^2}{2 + \cdots}}}},$$

which provides us with an example of a *continued fraction*, that is, a number expressed in the form

$$\alpha = a_0 + \cfrac{b_1}{a_1 + \cfrac{b_2}{a_2 + \cfrac{b_3}{a_3 + \cfrac{b_4}{a_4 + \cdots}}}},$$

where the a_i, b_i are integers. We shall restrict ourselves to *simple* continued fractions, which have the form

$$\alpha = a_0 + \cfrac{1}{a_1 + \cfrac{1}{a_2 + \cfrac{1}{a_3 + \cfrac{1}{a_4 + \cdots}}}},$$

where the a_i are positive integers.

We note that the expression can be finite or infinite; for example,

$$\frac{225}{157} = 1 + \cfrac{1}{2 + \cfrac{1}{3 + \cfrac{1}{4 + \cfrac{1}{5}}}} \quad \text{and} \quad \pi = 3 + \cfrac{1}{7 + \cfrac{1}{15 + \cfrac{1}{1 + \cfrac{1}{292 + \cdots}}}}.$$

Representation can be something of a typographical nightmare and, in consequence, more compact notations have been developed: principal among them is

$$[a_0; a_1, a_2, a_3, a_4, \ldots]$$

with the semi-colon used to separate the integer and the fractional parts of the number in question. In our two examples,

$$\frac{225}{157} = [1; 2, 3, 4, 5] \quad \text{and} \quad \pi = [3; 7, 15, 1, 292, \ldots].$$

The *convergents* of a continued fraction are the fractions

$$a_0 + \frac{1}{a_1}, \qquad a_0 + \cfrac{1}{a_1 + \cfrac{1}{a_2}}, \qquad a_0 + \cfrac{1}{a_1 + \cfrac{1}{a_2 + \cfrac{1}{a_3}}},$$

$$a_0 + \cfrac{1}{a_1 + \cfrac{1}{a_2 + \cfrac{1}{a_3 + \cfrac{1}{a_4}}}}, \qquad \text{etc.}$$

and produce progressively more accurate approximations to the number in question. In our cases,

$$\frac{225}{157} \sim \frac{3}{2}, \frac{10}{7}, \frac{43}{30} \quad \text{and} \quad \pi \sim 3, \frac{22}{7}, \frac{333}{106}, \frac{355}{113}, \frac{103993}{33102}.$$

We should note the appearance of that most famous rational approximation to π and the remarkable accuracy of the three subsequent approximations.

Manufacturing the convergents from the continued fraction is a matter of elementary arithmetic, as is the manufacture of the continued fraction itself from the number's decimal expansion, as we can see from

$$e = 2.718281828\ldots = 2 + 0.718281828\ldots$$

$$= 2 + \cfrac{1}{\left(\cfrac{1}{0.718281828\ldots}\right)} = 2 + \cfrac{1}{1.392211191\ldots}$$

$$= 2 + \cfrac{1}{1 + 0.392211191\ldots} = 2 + \cfrac{1}{1 + \cfrac{1}{\left(\cfrac{1}{0.392211191\ldots}\right)}}$$

$$= 2 + \cfrac{1}{1 + \cfrac{1}{2.549646778\ldots}} = 2 + \cfrac{1}{1 + \cfrac{1}{2 + 0.549646778\ldots}}$$

$$= 2 + \cfrac{1}{1 + \cfrac{1}{2 + \cfrac{1}{\left(\cfrac{1}{0.549646778\ldots}\right)}}}$$

$$= 2 + \cfrac{1}{1 + \cfrac{1}{2 + \cfrac{1}{1.819350244\ldots}}}, \quad \text{etc.,}$$

and we have the beginning of the continued fraction of

$$e = [2; 1, 2, 1, \ldots].$$

In the next section we will see Euler developing his ideas from this last calculation. In the section that follows matters take a devious turn, with the continued fraction representation not of a number, but of a function.

Euler and e

With his study of the limit

$$\lim_{n \to \infty} \left(1 + \frac{1}{n}\right)^n = e$$

it is probable that the great Jacob Bernoulli (1654–1705) should be credited with e's discovery but that the number is irrational must be attributed to a mathematician even greater than he. The irrationality was first recorded in a paper dated 1737 but published seven years later in the Proceedings of the National Academy of St. Petersburg.[1] Its Eneström number E71 reminds us that the work is located at that position in the universally accepted index of 866 entries, compiled by Gustav Eneström of the overwhelmingly prolific eighteenth-century Swiss genius Leonhard Euler (1707–83). Euler had turned his ever-restless, ever-incisive, all-inclusive mathematical mind to the matter of continued fractions, recording his deep thoughts on the matter for the first time (although he was to return to the subject often throughout his long and incomparably impressive career[2]). The paper's title is *De fractionibus continuis dissertation: An essay on continued fractions.* In the preamble he commented[3]:

> Since I have been studying continued fractions for a long time, and since I have observed many important facts pertaining both to their use and their derivation, I have decided to discuss them here. Although I have not yet arrived at a complete theory, I believe that these partial results which I have found after hard work will surely contribute to further study of this subject.

In the course of its forty pages Euler touches on some basic algebraic theory as well as the far more subtle analytic implications associated with continued fractions, with the degree of *hard work* undertaken by him quite evident, even for someone of his remarkable calculative abilities. The paper charts a significant area of the early theory of continued fractions and we will choose a path through it that is most convenient for our purpose, which begins with an observation made in Section 11a:

> The transformation of an ordinary fraction into a continued fraction with numerators all 1 and integral denominators must first be shown. Moreover, every finite fraction whose

---

[1]Presented to the Academy on March 7, 1737.

[2]The interested reader may wish to consult *The Euler Archive*, http://math.dartmouth.edu/~euler/.

[3]Translation by M. F. Wyman and B. F. Wyman, 1985, *Mathematical Systems Theory* 18:295–328.

> numerators and denominators are finite whole numbers may
> be transformed into a continued fraction of this kind which
> is truncated at a finite level. On the other hand, a fraction
> whose numerator and denominator are infinitely large num-
> bers (which are given for irrational and transcendental quan-
> tities) will go across to a continued fraction running to infin-
> ity. To find such a continued fraction, it suffices to assign the
> denominators, since we set all the numerators equal to 1.

Put in other words, Euler knew that the simple continued fraction
form of a rational number is finite and that of an irrational number
unending. To prove e irrational he needed to manufacture such a
continued fraction: the stage is set for one of his astonishing and
penetrating manipulations.

We move to sections 21 and 22 of the paper we see him turning
his attention to the continued fraction form of his approximation
to e ~ 2.71828182845904. These 14 decimal places would have
required him to evaluate the sum $\sum_{r=0}^{16} 1/r!$ for the known infinite
series expansion for e and then to perform repeatedly the process
for generating the continued fraction

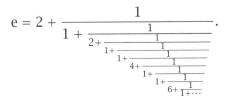

Add to this the effort required of him as he experiments with numbers related to e, as we show below, and the reader will begin to
have some measure of the immense calculative ability of the man.
To quote the distinguished eighteenth-century French politician
and mathematician Dominique François Jean Arago:

> Euler calculated without effort, just as men breathe, as eagles
> sustain themselves in the air.

The regular pattern of the denominators hardly escaped him and
he noted that the

> third denominators make up the arithmetic progression 2, 4,
> 6, 8, etc., the others being ones. Even if this rule is seen by
> observation alone, nevertheless it is reasonable to suppose that
> it extends to infinity. In fact, this result will be proved below.

Of course, such a proof would render e to be irrational.

His relentless desire to experiment had him move to $\sqrt{e} = 1.6487212707$ and the continued fraction

$$\sqrt{e} = 1 + \cfrac{1}{1 + \cfrac{1}{1 + \cfrac{1}{1 + \cfrac{1}{5 + \cfrac{1}{1 + \cfrac{1}{1 + \cfrac{1}{9 + \cfrac{1}{1 + \cfrac{1}{1 + \cfrac{1}{13 + \cdots}}}}}}}}}}.$$

And once again, what he describes as the *interrupted* arithmetic progression, is noted. These interruptions are seen to disappear, apart from a disingenuous first term, with

$$\frac{\sqrt[3]{e} - 1}{2} = 0.1978062125\ldots = 1 + \cfrac{1}{5 + \cfrac{1}{18 + \cfrac{1}{30 + \cfrac{1}{42 + \cfrac{1}{54 + \cdots}}}}}.$$

And to disappear altogether with

$$\frac{e^2 - 1}{2} = 3.19452804951\ldots = 3 + \cfrac{1}{5 + \cfrac{1}{7 + \cfrac{1}{9 + \cfrac{1}{11 + \cfrac{1}{13 + \cfrac{1}{15 + \cdots}}}}}},$$

He had not reached but was homing in to the study of continued fractions having numerators 1 and denominators in arithmetic progression, which he characterized in Section 31 as

$$s = a + \cfrac{1}{(1 + n)a + \cfrac{1}{(1 + 2n)a + \cfrac{1}{(1 + 3n)a + \cfrac{1}{(1 + 4n)a + \cdots}}}}.$$

Commenting:

> Furthermore, from the number e itself, whose continued fraction has an interrupted arithmetic progression of denominators, I have observed that with a few changes of this kind a

continued fraction free of interruptions can be formed, For
example,

$$\frac{e+1}{e-1} = 2 + \cfrac{1}{6 + \cfrac{1}{10 + \cfrac{1}{14 + \cfrac{1}{18 + \cfrac{1}{22 + \cfrac{1}{26 + \cdots}}}}}}.$$

Now he had manufactured an 'uninterrupted' arithmetic progression and one which is a special case of his general schema, with
$a = n = 2$. (He gave no explicit decimal approximation for the
number.)

> In the preceding sections, where I have converted the number
> e (whose logarithm is 1) together with its powers into contin-
> ued fractions, I have only observed the arithmetic progression
> of the denominators and I have not been able to affirm any-
> thing except the probability of this progression continuing to
> infinity. Therefore, I have exerted myself in this above all: that
> I might inquire into the necessity of this progression and prove
> it rigorously. Even this goal I have pursued in a peculiar way...

Euler's investigation into the nature of his general arithmetic progression type of continued fraction begins with him writing out
the first few convergents:

$$a, \quad \frac{(1+n)a^2 + 1}{(1+n)a}, \quad \frac{(1+n)(1+2n)a^3 + (2+2n)a}{(1+n)(1+2n)a^2 + 1}, \quad \cdots$$

and then, using arguments from earlier in the paper, rewrites the
continued fraction as the ratio of two power series

$$s = \frac{p}{q},$$

where the variables p and q are defined as

$$p = a + \frac{1}{1.na} + \frac{1}{1.2.1(1+n)n^2 a^3}$$

$$+ \frac{1}{1.2.3.1(1+n)(1+2n)n^3 a^5} + \cdots$$

and

$$q = 1 + \frac{1}{1(1+n)na^2} + \frac{1}{1.2(1+n)(1+2n)n^2a^4}$$
$$+ \frac{1}{1.2.3(1+n)(1+2n)(1+3n)n^3a^6} + \cdots.$$

Then follows a series of remarkable changes of variable. First, the variable z is defined by $a = 1/\sqrt{nz}$ to give

$$s = \frac{1}{\sqrt{nz}} \frac{t}{u},$$

where the variables t and u are defined as

$$t = 1 + \frac{z}{1.1} + \frac{z^2}{1.2.1(1+n)} + \frac{z^3}{1.2.3.1(1+n)(1+2n)} + \cdots$$

and

$$u = 1 + \frac{z}{1(1+n)} + \frac{z^2}{1.2(1+n)(1+2n)}$$
$$+ \frac{z^3}{1.2.3(1+n)(1+2n)(1+3n)} + \cdots.$$

He noted that $dt/dz = u$ and also that

$$t - u = \left(\frac{nz}{1(1+n)} + \frac{nz^2}{1.(1+n)(1+2n)} \right.$$
$$\left. + \frac{nz^3}{1.2(1+n)(1+2n)(1+3n)} + \cdots \right)$$
$$= nz\frac{du}{dz}.$$

We then have

$$nz\frac{du}{dz} = t - u.$$

The next substitution is $t = uv$, thereby defining the variable v and making $s = v/\sqrt{nz}$.

Using the product and chain rules,

$$u = \frac{dt}{dz} = \frac{d(uv)}{dz} = u\frac{dv}{dz} + v\frac{du}{dz},$$

which makes

$$u\frac{dv}{dz} = u - v\frac{du}{dz}.$$

Now,

$$nz\frac{du}{dz} = t - u = uv - u,$$

so

$$\frac{du}{dz} = \frac{uv - u}{nz}.$$

Substituting from above gives

$$u\frac{dv}{dz} = u - v\frac{uv - u}{nz}$$

or

$$\frac{dv}{dz} = 1 - v\frac{v - 1}{nz},$$

$$nz\frac{dv}{dz} = nz - v^2 + v,$$

$$nz\frac{dv}{dz} + v^2 - v = nz.$$

Finally, he defined r and q by $z = r^n$ and $v = qr$ and engaged in yet more nifty calculus, which in modern form is

$$\frac{dv}{dz} = \frac{d(qr)}{dz} = q\frac{dr}{dz} + r\frac{dq}{dz} = q\frac{dr}{dz} + r\frac{dq}{dr}\frac{dr}{dz}$$

$$= \left(q + r\frac{dq}{dr}\right)\frac{dr}{dz},$$

$$nz\frac{dv}{dz} = nz\left(q + r\frac{dq}{dr}\right)\frac{dr}{dz} = nr^n\left(q + r\frac{dq}{dr}\right)\frac{1}{nr^{n-1}}$$

$$= r\left(q + r\frac{dq}{dr}\right),$$

which makes

$$nz\frac{dv}{dz} + v^2 - v = r\left(q + r\frac{dq}{dr}\right) + q^2r^2 - qr$$

$$= r^2\frac{dq}{dr} + q^2r^2 = nz = nr^n,$$

Table 3.1.

Expression	Defined variable
$a = 1/\sqrt{nz}$	z
t = numerator	t
u = denominator	u
$t = uv$	v
$z = r^n$	r
$v = qr$	q

which simplifies to

$$r^2\frac{\mathrm{d}q}{\mathrm{d}r} + q^2r^2 = nr^n,$$

$$\frac{\mathrm{d}q}{\mathrm{d}r} + q^2 = nr^{n-2},$$

from which he observed:

> From this equation, if q may be determined from r and r is set equal to $r = n^{-1/n}a^{-2/n}$ then the desired value will be $s = aqr$.

And he comments that the differential equation

> agrees with the equation once proposed by Count Riccati.

At this point the reader might benefit from a summary of the defined variables and the equations defining them, which is provided in table 3.1. In which case, as Euler observed, $s = v/\sqrt{nz} = av = aqr$.

Euler had been working on a particularly difficult type of differential equation,[4] which had been proposed in 1720 by the Venetian nobleman Count Jacopo Francesco Riccati (1676–1754) and somehow he had managed to transform his problem of infinite continued fractions to its solution; quite how he matched the two, we are left to wonder. He returned both to continued fractions and to the further delicate study of the Riccati Equation and to

[4]E70, *De constructione aequationum.* Presented to the St. Petersburg Academy on February 7, 1737.

the combination of both of them[5] but fortunately the situation of importance to us is $n = a = 2$, a special case which makes the solution of the equation easy. So,

$$\frac{dq}{dr} + q^2 = 2, \quad \int \frac{dq}{q^2 - 2} = -\int dr, \quad \frac{1}{2\sqrt{2}} \ln \frac{q + \sqrt{2}}{q - \sqrt{2}} = r + c,$$

which makes

$$q = \sqrt{2} \frac{Ae^{2\sqrt{2}r} + 1}{Ae^{2\sqrt{2}r} - 1}.$$

And to make $v = qr$ always finite we have his boundary condition '$q = \infty, r = 0$' to give $A = 1$ and

$$q = \sqrt{2} \frac{e^{2\sqrt{2}r} + 1}{e^{2\sqrt{2}r} - 1}.$$

And finally

$$r = \frac{1}{a\sqrt{2}} \quad \text{and} \quad s = arq = \frac{q}{\sqrt{2}}, \quad \text{which makes} \quad s = \frac{e^{2/a} + 1}{e^{2/a} - 1}.$$

Using the fact that, in our case, $a = 2$, we have

$$\frac{e + 1}{e - 1} = 2 + \cfrac{1}{6 + \cfrac{1}{10 + \cfrac{1}{14 + \cfrac{1}{18 + \cfrac{1}{22 + \cfrac{1}{26 + \cdots}}}}}}.$$

The number $(e + 1)/(e - 1)$ has an infinite arithmetic progression as its denominators in its representation as a continued fraction: therefore $(e+1)/(e-1)$ is irrational, therefore e is irrational. There is no such triumphant statement in E71, merely a confirmation that earlier empirical suspicions have been confirmed.

Of course, with his continued fraction form for $(e^2 - 1)/2$ there is effectively nothing to do to prove that it and therefore e^2 is irrational; a stronger result still.

---

[5] In a paper dated 20 March 1780 and entitled E751 – Analysis facilis aequationem Riccatianam per fractionem continuam resolvendi, but which was originally published in *Mémoires de l'académie des sciences de St.-Petersbourg* 6:12–29 (1818).

Lambert and π

Even under Eulerian scrutiny the simple continued fraction representation of π

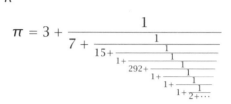

$$\pi = 3 + \cfrac{1}{7 + \cfrac{1}{15 + \cfrac{1}{1 + \cfrac{1}{292 + \cfrac{1}{1 + \cfrac{1}{1 + \cfrac{1}{1 + \cfrac{1}{2 + \cdots}}}}}}}}$$

offered little encouragement in establishing the number's irrationality. The nice pattern with which Euler juggled is absent, yet continued fractions do have within them the means to do the job and it was a contemporary of his, younger by 21 years, who managed (at the cost of yet more considerable effort) to be first to publish a proof. Johann Heinrich Lambert (1728–77), philosopher, physicist, geometer, probabilist and number theorist, once acolyte and friend of Euler, later antagonist to him, was to die at the age of 49. It seems remarkable today that the friendship of these two remarkable compatriots of the Berlin Academy should have been irrevocably damaged by disagreements which began with differing views on the Academy's sale of calendars.[6] Yet Lambert's comparatively short life was spectacularly fruitful, with his 1761 proof[7] that π is irrational one of its mathematical highpoints. To be more exact, he proved that, if $x \neq 0$ is rational, then $\tan(x)$ is irrational: using the logical reverse of this statement, since $\tan(\pi/4) = 1$ is rational, it must be that $\pi/4$ and so π is irrational.[8]

His proof uses a continued fraction representation not of a number but of a function, $\tan(x)$ (where x is measured in radians) and may be separated into two parts.

---

[6]N. N. Bogolyubov, G. K. Mikhailov and A. P. Yushkevich (eds), 2007, *Euler and Modern Science* (Mathematical Association of America).

[7]J. H. Lambert, 1761, Mémoire sur quelques propriétés remarquables des quantités transcendentes circulaires et logarithmiques, *Histoire de l'Académie Royale des Sciences et des Belles-Lettres der Berlin* 17:265–322. Reprinted in 1948 in *Iohannis Henrici Lambert, Opera Mathematica*, Vol. II, ed. A. Speiser (Zürich: Orell Füssli).

[8]In the same paper and by the same means he also proved the same result for e^x, using complex numbers and the continued fraction form of the function $\tanh(x)$.

First, he needed to establish the continued fraction form of

$$\tan x = \frac{\sin x}{\cos x}$$

and did so by starting with the series expansions of

$$\sin x = x - \frac{x^3}{3!} + \frac{x^5}{5!} - \frac{x^7}{7!} + \cdots$$

and

$$\cos x = 1 - \frac{x^2}{2!} + \frac{x^4}{4!} - \frac{x^6}{6!} + \cdots .$$

He had, then,

$$\tan x = \frac{x - \dfrac{x^3}{3!} + \dfrac{x^5}{5!} - \dfrac{x^7}{7!} + \cdots}{1 - \dfrac{x^2}{2!} + \dfrac{x^4}{4!} - \dfrac{x^6}{6!} + \cdots}.$$

The move from the ratio of two power series to the infinite continued fraction (the reverse of Euler's approach described earlier) was achieved by a sequence of horrendous manipulations which are effectively a mixture of long division and recursion; they serve no purpose here, but the interested (patient and brave) reader is of course at liberty to pursue the details.[9] This omission cuts a considerable swathe through Lambert's involved and ingenious argument but we think the reader will judge that what follows is sufficient to impress! He had established that

$$\tan x = \cfrac{x}{1 - \cfrac{x^2}{3 - \cfrac{x^2}{5 - \cfrac{x^2}{7 - \cfrac{x^2}{9 - \cdots}}}}} = \cfrac{x}{1 + \cfrac{-x^2}{3 + \cfrac{-x^2}{5 + \cfrac{-x^2}{7 + \cfrac{-x^2}{9 + \cdots}}}}}.$$

[9]Pierre Eymard and Jean-Pierre Lafon, 2004, *The Number π* (American Mathematical Society).

The second part of the proof relies on the following two properties of continued fractions.

For the continued fraction

$$y = b_0 + \cfrac{a_1}{b_1 + \cfrac{a_2}{b_2 + \cfrac{a_3}{b_3 + \cfrac{a_4}{b_4 + \cdots}}}}.$$

1. If $\{\lambda_1, \lambda_2, \lambda_3, \dots\}$ is an infinite sequence of non-zero numbers, the continued fraction

$$b_0 + \cfrac{\lambda_1 a_1}{\lambda_1 b_1 + \cfrac{\lambda_1 \lambda_2 a_2}{\lambda_2 b_2 + \cfrac{\lambda_2 \lambda_3 a_3}{\lambda_3 b_3 + \cfrac{\lambda_3 \lambda_4 a_4}{\lambda_4 b_4 + \cdots}}}}$$

has the same convergents as y and, if there is convergence, converges to y.

2. Now suppose that $b_0 = 0$.

If $|a_i| < |b_i|$ for all $i \geqslant 1$, then:

(a) $|y| \leqslant 1$;

(b) writing

$$y_n = \cfrac{a_n}{b_n + \cfrac{a_{n+1}}{b_{n+1} + \cfrac{a_{n+2}}{b_{n+2} + \cfrac{a_{n+3}}{b_{n+3} + \cdots}}}}$$

if, for some n and above, it is never the case that $|y_n| = 1$, then y is irrational.

We shall not prove these here, but the interested reader is referred to Appendix C on page 281.

Now consider that continued fraction expansion for $y = \tan x$ and suppose that $x = p/q$. Then

$$\tan\left(\frac{p}{q}\right) = \cfrac{p/q}{1 + \cfrac{-p^2/q^2}{3 + \cfrac{-p^2/q^2}{5 + \cfrac{-p^2/q^2}{7 + \cfrac{-p^2/q^2}{9 + \cdots}}}}}.$$

And we can eliminate the denominators by using property 1 above with $\lambda_i = q$ for all i to get

$$\tan\left(\frac{p}{q}\right) = \cfrac{p}{q + \cfrac{-p^2}{3q + \cfrac{-p^2}{5q + \cfrac{-p^2}{7q + \cfrac{-p^2}{9q + \cdots}}}}}.$$

Now define y_n in the manner above:

$$y_n = \cfrac{-p^2}{(2n+1)q + \cfrac{-p^2}{(2n+3)q + \cfrac{-p^2}{(2n+5)q + \cdots}}}$$

but ensure that n is so big that $p^2 < 2nq$. This means that $|-p^2| < (2n+r)q$ for $r = 1, 3, 5, \ldots$ and this means that condition 2 above is satisfied and so $|y_n| \leqslant 1$; evidently, $|y_{n+1}| \leqslant 1$ also.

But $y_n = -p^2/((2n+1)q + y_{n+1})$, where the numerator is evidently negative and the denominator $(2n+1)q + y_{n+1} = 2nq + (q + y_{n+1})$ positive since $q \geqslant 1$ and $|y_{n+1}| \leqslant 1$, which means that

$$|y_n| = \left| \frac{-p^2}{(2n+1)q + y_{n+1}} \right| = \frac{p^2}{2nq + (q + y_{n+1})} \leqslant \frac{p^2}{2nq} < 1.$$

So, although $|y_n| \leqslant 1$ we have $|y_n| < 1$ and so $y_n \neq \pm 1$ for all sufficiently large n. This means that $\tan(p/q)$ is irrational. As the man himself said,

> The diameter (of a circle) does not stand to the circumference
> as an integer to an integer.

And with these arguments we have followed in the footsteps of mathematical pioneers of the highest calibre. We have already commented that 'first' is seldom associated with 'pretty' in mathematical proof and it is often not associated with 'rigorous' either, particularly in earlier days. In the next chapter we will move on to more modern arguments that establish these two results and many more besides.

Irrationals, Old and New

> The irrationals exist in such variety, indeed, that no notation whatever is capable of providing a separate name for each of them
>
> Willard Van Orman Quine

If the purpose of the previous chapter was simple, that of the present one is assuredly not so. We will go in search of irrational numbers: all manner of them. In doing so we will begin with more recent approaches which establish the irrationality of e and π, move on to more general methods which establish much more, approach new types of irrationals and finish with transcendentals.

Fourier and e

It is strange that a mathematician of Euler's stature had not detected that the irrationality of e was an inevitable consequence of its canonical representation as the infinite series, well known to him:

$$e = 1 + \frac{1}{1!} + \frac{1}{2!} + \frac{1}{3!} + \frac{1}{4!} + \cdots .$$

It would take until 1815 for the series to be used to establish the number's irrationality and then by Joseph Fourier, who is better known today for his theory of heat conduction. This said, there is no mention of the result in Fourier's extant writings, and we must rely on a footnote to paragraph 232 on page 341 of a voluminous tome of one M. J. de Stainville entitled *Mélanges d'Analyse Algébrique et de Géometrie* for evidence. The paragraph contains a version of the proof, at the end of which is written

> Cette démonstration m'a été communiquée par M. Poinsot, qui m'a dit la tenir de M. Fourier.[1]

de Stainville, a Parisian mathematics teacher who was to end his life in an asylum, was in contact with his distinguished contemporary Louis Poinsot, a significant geometer and the inventor of geometrical mechanics, which challenges the modern high school students to resolve forces. We have, then, the Fourier connection, and with the book's publication date of 1815, we have a date.

The proof is, as ever, by contradiction. Since e is easily shown to lie between two integers, it cannot itself be an integer and if we assume it to be rational we may write e = m/n, with $n > 1$, and so separate the series as

$$e = \left(1 + \frac{1}{1!} + \frac{1}{2!} + \frac{1}{3!} + \frac{1}{4!} + \cdots + \frac{1}{n!}\right) + \cdots.$$

Then

$$n!e = n!\frac{m}{n} = (n-1)!m$$
$$= \left(n! + \frac{n!}{1!} + \frac{n!}{2!} + \frac{n!}{3!} + \frac{n!}{4!} + \cdots + \frac{n!}{n!}\right) + \cdots + R,$$

which makes $R \neq 0$ the difference between two integers, and so an integer itself.

But

$$R = n!\left(\frac{1}{(n+1)!} + \frac{1}{(n+2)!} + \frac{1}{(n+3)!} + \cdots\right)$$
$$= \frac{1}{n+1} + \frac{1}{(n+1)(n+2)} + \frac{1}{(n+1)(n+2)(n+3)} + \cdots$$
$$< \frac{1}{n+1} + \frac{1}{(n+1)^2} + \frac{1}{(n+1)^3} + \cdots$$
$$= \left(\frac{1}{n+1}\right)\Big/\left(1 - \frac{1}{n+1}\right) = \frac{1}{n} < 1.$$

Our (strictly positive) integer R is less than 1 and we have our contradiction.

---

[1] This demonstration was communicated to be by M. Poinsot, who attributed it to M. Fourier.

In fact the argument allows an extension which discloses the irrationality of e^2.

Again suppose that $e^2 = M/N$, $N > 1$, then $Ne = M \times 1/e$ and this means that

$$N\left(1 + \frac{1}{1!} + \frac{1}{2!} + \frac{1}{3!} + \cdots + \frac{1}{n!} + \cdots\right)$$

$$= M\left(1 - \frac{1}{1!} + \frac{1}{2!} - \frac{1}{3!} + \cdots + \frac{(-1)^n}{n!} \cdots\right).$$

For any choice of n, multiply both sides by $n!$ to get

$$N\left(I_n + \frac{n!}{(n+1)!} + \frac{n!}{(n+2)!} + \frac{n!}{(n+3)!} + \cdots\right)$$

$$= M\left(J_n + (-1)^{n+1}\left(\frac{n!}{(n+1)!} - \frac{n!}{(n+2)!} + \frac{n!}{(n+3)!} - \cdots\right)\right),$$

where I_n, J_n are integers, as described above.

This makes

$$NI_n + N\left(\frac{1}{n+1} + \frac{1}{(n+1)(n+2)} + \frac{1}{(n+1)(n+2)(n+3)} + \cdots\right)$$

$$= MJ_n + M(-1)^{n+1}\left(\frac{1}{n+1} - \frac{1}{(n+1)(n+2)}\right.$$

$$\left. + \frac{1}{(n+1)(n+2)(n+3)} - \cdots\right),$$

which we rearrange as

$$NI_n - MJ_n = M(-1)^{n+1}\left(\frac{1}{n+1} - \frac{1}{(n+1)(n+2)}\right.$$

$$\left. + \frac{1}{(n+1)(n+2)(n+3)} - \cdots\right)$$

$$- N\left(\frac{1}{n+1} + \frac{1}{(n+1)(n+2)}\right.$$

$$\left. + \frac{1}{(n+1)(n+2)(n+3)} + \cdots\right)$$

and so, using the triangle inequality,

$$|NI_n - MJ_n|$$

$$\leqslant M \left| \frac{1}{n+1} - \frac{1}{(n+1)(n+2)} + \frac{1}{(n+1)(n+2)(n+3)} - \cdots \right|$$

$$+ N \left| \frac{1}{n+1} + \frac{1}{(n+1)(n+2)} + \frac{1}{(n+1)(n+2)(n+3)} + \cdots \right|$$

$$< \frac{M+N}{n}$$

since we already know that both infinite sums are bounded above by $1/n$. Now make n large enough to make the positive integer on the left less than 1, and so 0: for this infinite set of n, $NI_n = MJ_n$, which is clearly impossible.

The appearance of e as an infinite series has allowed a simple and effective argument to be put to establish its irrationality (as well as that of e^2 (the arguments can be further extended too)) and there are myriad other provably irrational constants defined by infinite series, but the reader should curb any optimism that a series form necessarily leads to a proof of irrationality. Taking three examples (of many), the nature of the following numbers remains a mystery:

• Relating to e,

$$\sum_{k=0}^{\infty} \frac{1}{k!+1} = 1.52606813447333\ldots,$$

  asked by Paul Erdős.
• The Catalan constant

$$G = \sum_{k=1}^{\infty} \frac{(-1)^{k-1}}{(2k-1)^2} = 0.915965594177\ldots.$$

• The Euler–Mascheroni constant

$$\gamma = \lim_{n \to \infty} \left(\sum_{k=1}^{n} \frac{1}{k} - \ln n \right) = 0.577215664901532\ldots.$$

Hermite and π

A typically demanding problem in the Cambridge University Mathematics Preliminary Examination was one set in 1945 by the eminent mathematician Dame Mary Cartwright, culled from her past experience. It is a structured question on the proof of the irrationality not of π but of the stronger result of π^2; of the origin of the methods involved she claimed uncertainty, but of central importance is the observation that

$$\frac{1}{n!}\left(\frac{\pi}{2}\right)^{2n+1}\int_{-1}^{1}(1-x^2)^n\cos(\tfrac{1}{2}\pi x)\,dx$$

is a polynomial in $\pi^2/4$ with integer coefficients and degree $\lfloor n/2 \rfloor$ (where "$\lfloor \cdot \rfloor$" is the Floor function) and that, from the assumption that π^2 is rational, it is again possible to deduce the existence of a positive integer less than 1. In fact, the integral is to be found in the work of the French nineteenth-century mathematician Charles Hermite (1822–1901),[2] whose contribution to the story of irrational numbers is more significant still, as we will see later in this chapter and also in the next. For now we will be content to look in detail at that examination problem which uses his integral to establish the irrationality of π^2.

Write

$$I_n = \int_{-1}^{1}(1-x^2)^n\cos(\tfrac{1}{2}\pi x)\,dx, \quad n = 0,1,2,\dots.$$

First notice that, for all such n, the integral is bounded, since

$$0 < (1-x^2)^n\cos(\tfrac{1}{2}\pi x) < 1 \quad \text{for } -1 < x < 1$$

and so

$$\int_{-1}^{1}0\,dx < \int_{-1}^{1}(1-x^2)^n\cos(\tfrac{1}{2}\pi x)\,dx < \int_{-1}^{1}1\,dx,$$

which makes $0 < I_n < 2$. Inevitably, we need to establish the nature of the integral and to this end we shall evaluate a few

---

[2] C. Hermite, 1873, Extrait d'une lettre de Mr. Charles Hermite à Mr. Borchardt, *J. De Crelle* 76:342–44; *Oeuvres*, t. III, 1912, pp. 146–49 (Gauthier-Villars, Paris).

examples of it or, to be precise, have a mathematical computer package evaluate them for us:

$$(\tfrac{1}{2}\pi)^{2\times0+1}I_0 = 2,$$

$$(\tfrac{1}{2}\pi)^{2\times1+1}I_1 = 4,$$

$$(\tfrac{1}{2}\pi)^{2\times2+1}I_2 = 2!(96 - 8(\tfrac{1}{4}\pi^2)),$$

$$(\tfrac{1}{2}\pi)^{2\times3+1}I_3 = 3!(960 - 96(\tfrac{1}{4}\pi^2)),$$

$$(\tfrac{1}{2}\pi)^{2\times4+1}I_4 = 4!(3360 - 1440(\tfrac{1}{4}\pi^2) + 32(\tfrac{1}{4}\pi^2)^2),$$

$$(\tfrac{1}{2}\pi)^{2\times5+1}I_5 = 5!(60480 - 26880(\tfrac{1}{4}\pi^2) + 960(\tfrac{1}{4}\pi^2)^2).$$

The results for a given n can be interpreted both as polynomials with integer coefficients of degree $2\lfloor\tfrac{1}{2}n\rfloor$ in $\tfrac{1}{2}\pi$ or, as they have been written, polynomials with integer coefficients of degree $\lfloor\tfrac{1}{2}n\rfloor$ in $\tfrac{1}{4}\pi^2$. Taking the latter interpretation we can approach a proof that π^2 is irrational.

So, the suggestion is that

$$(\tfrac{1}{2}\pi)^{2n+1}I_n = n!P_n, \qquad\qquad (*)$$

where P_n is a polynomial with integral coefficients (depending on n) in $\tfrac{1}{4}\pi^2$ of degree $\lfloor\tfrac{1}{2}n\rfloor$.

With such a suspicion, a proof by induction is a natural way to progress and to that end we need a recurrence formula for the I_n, with integration by parts twice providing it:

$$
\begin{aligned}
I_n &= \int_{-1}^{1} (1 - x^2)^n \cos(\tfrac{1}{2}\pi x)\, dx \\
&= \left[\frac{2}{\pi}(1 - x^2)^n \sin(\tfrac{1}{2}\pi x) \right]_{-1}^{1} \\
&\quad + \frac{4n}{\pi} \int_{-1}^{1} x(1 - x^2)^{n-1} \sin(\tfrac{1}{2}\pi x)\, dx \\
&= \frac{4n}{\pi} \int_{-1}^{1} x(1 - x^2)^{n-1} \sin(\tfrac{1}{2}\pi x)\, dx \\
&= \frac{4n}{\pi} \left\{ \left[-\frac{2}{\pi}x(1 - x^2)^{n-1} \cos(\tfrac{1}{2}\pi x) \right]_{-1}^{1} \right. \\
&\quad + \frac{2}{\pi} \int_{-1}^{1} \{(1 - x^2)^{n-1} - 2(n - 1)x^2(1 - x^2)^{n-2}\} \\
&\quad\quad\quad\quad\quad\quad\quad\quad\quad\quad\quad\quad \left. \times \cos(\tfrac{1}{2}\pi x)\, dx \right\}
\end{aligned}
$$

$$= \frac{4n}{\pi} \left\{ \frac{2}{\pi} \int_{-1}^{1} \{(1-x^2)^{n-1} - 2(n-1)x^2(1-x^2)^{n-2}\} \right.$$
$$\left. \times \cos(\tfrac{1}{2}\pi x)\, dx \right\}$$

$$= \frac{8n}{\pi^2} \left\{ \int_{-1}^{1} \{(1-x^2)^{n-1} - 2(n-1)x^2(1-x^2)^{n-2}\} \right.$$
$$\left. \times \cos(\tfrac{1}{2}\pi x)\, dx \right\}.$$

Now rewrite the expression

$$I_n = \frac{8n}{\pi^2} \left\{ \int_{-1}^{1} \{(1-x^2)^{n-1} + 2(n-1)((1-x^2)-1)(1-x^2)^{n-2}\} \right.$$
$$\left. \times \cos(\tfrac{1}{2}\pi x)\, dx \right\}$$

$$= \frac{8n}{\pi^2}(I_{n-1} + 2(n-1)I_{n-1} - 2(n-1)I_{n-2})$$

and so

$$\tfrac{1}{4}\pi^2 I_n = 2n(2n-1)I_{n-1} - 4n(n-1)I_{n-2}, \quad n \geqslant 2.$$

With this in place we take the inductive step, with the assumption that condition $(*)$ is true for all integers $\leqslant k$, and consider the situation for $k+1$. Using the recurrence relation we have

$$(\tfrac{1}{2}\pi)^{2k+3}I_{k+1}$$
$$= (\tfrac{1}{2}\pi)^{2k+3}\left(\frac{2}{\pi}\right)^2 \{2(k+1)(2k+1)I_k - 4(k+1)kI_{k-1}\}$$
$$= (\tfrac{1}{2}\pi)^{2k+1}\{2(k+1)(2k+1)I_k - 4(k+1)kI_{k-1}\}$$
$$= 2(k+1)(2k+1)(\tfrac{1}{2}\pi)^{2k+1}I_k - 4(k+1)k(\tfrac{1}{2}\pi)^2(\tfrac{1}{2}\pi)^{2k-1}I_{k-1}$$
$$= 2(k+1)(2k+1)k!P_k - 4(k+1)k(\tfrac{1}{4}\pi^2)(k-1)!P_{k-1}.$$

And we have the polynomial in $\tfrac{1}{4}\pi^2$ with integer coefficients and of degree

$$\lfloor \tfrac{1}{2}(k-1)\rfloor + 1 = \lfloor \tfrac{1}{2}(k-1)+1\rfloor = \lfloor \tfrac{1}{2}(k+1)\rfloor.$$

We already have the assumption true for some starting values and so we are done and we have that $(\tfrac{1}{2}\pi)^{2n+1}I_n = n!P_n$.

Finally, suppose that $\frac{1}{4}\pi^2 = a/b$ and square both sides of the result to give

$$(\tfrac{1}{4}\pi^2)^{2n+1}I_n^2 = (n!)^2(P_n)^2$$

and so

$$\left(\frac{a}{b}\right)^{2n+1} I_n^2 = (n!)^2(P_n)^2,$$

which means that

$$\frac{a^{2n+1}}{(n!)^2}I_n^2 = b^{2n+1}(P_n)^2.$$

Since P_n is a polynomial in a/b with integer coefficient and of degree $\lfloor \frac{1}{2}n \rfloor$, $(P_n)^2$ has degree $2\lfloor \frac{1}{2}n \rfloor$ and so it must be that $b^{2n+1}P_n$ is an integer. But $a^{2n+1}/(n!)^2 \xrightarrow{n \to \infty} 0$, hence for sufficiently large n, since I_n is bounded, as previously mentioned, $0 < a^{2n+1}I_n/n! < 1$.

And we have that integer between 0 and 1.

Niven and Others

The Carus Mathematical Monograph series of the Mathematical Association of America has as its 11th volume *Irrational Numbers*,[3] authored in 1955 by the acclaimed Canadian–American mathematician Ivan Niven (1915-99). Among numerous results connected with irrational numbers it contains his version of a proof of the irrationality of π, adapted from his original note of 1947,[4] which was the first modern proof of π's irrationality to appear in print.

For this proof, and for much else in the book, he adopted a device somewhat reminiscent of Hermite's integral, which has at its heart a polynomial of degree $2n$, the canonical form of which is

$$f(x) = \frac{x^n(1-x)^n}{n!}.$$

Its common name is a *Niven polynomial*.

---

[3]Ivan Niven, 2005, *Irrational Numbers*, Carus Mathematical Monographs (Mathematical Association of America).

[4]Ivan Niven, 1947, A simple proof that π is irrational, *Bull. Am. Math. Soc.* 53:509.

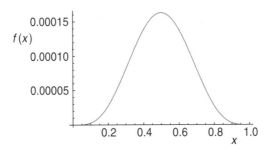

Figure 4.1.

Of course, its nature varies as n changes but if we restrict ourselves to the interval $[0, 1]$ its visual behaviour is typified for any
positive integer n and we illustrate matters in figure 4.1 with the
case $n = 4$.

Essential aspects of its behaviour remain constant as n varies
too. Again, let us deal with the case $n = 4$ and adopt the standard
notation $f^{(r)}(x)$ for $(\mathrm{d}^r/\mathrm{d}x^r)f(x)$:

$$f(x) = \frac{x^4(1-x)^4}{4!}: \quad f(0) = f(1) = 0,$$

$$f^{(1)}(x) = \tfrac{1}{6}(1-x)^4 x^3 - \tfrac{1}{6}(1-x)^3 x^4: \quad f^{(1)}(0) = f^{(1)}(1) = 0,$$

$$f^{(2)}(x) = \tfrac{1}{2}(1-x)^4 x^2 - \tfrac{4}{3}(1-x)^3 x^3 + \tfrac{1}{2}(1-x)^2 x^4:$$
$$f^{(2)}(0) = f^{(2)}(1) = 0,$$

$$f^{(3)}(x) = (1-x)^4 x - 6(1-x)^3 x^2 + 6(1-x)^2 x^3 - (1-x)x^4:$$
$$f^{(3)}(0) = f^{(3)}(1) = 0,$$

$$f^{(4)}(x) = (1-x)^4 - 16(1-x)^3 x + 36(1-x)^2 x^2$$
$$- 16(1-x)x^3 + x^4: f^{(4)}(0) = f^{(4)}(1) = 1,$$

$$f^{(5)}(x) = -20(1-x)^3 + 120(1-x)^2 x - 120(1-x)x^2 + 20x^3:$$
$$f^{(5)}(0) = -20 = -f^{(5)}(1),$$

$$f^{(6)}(x) = 180(1-x)^2 - 480(1-x)x + 180x^2:$$
$$f^{(6)}(0) = f^{(6)}(1) = 180,$$

$$f^{(7)}(x) = -840(1-x) + 840x: \quad f^{(7)}(0) = -840 = -f^{(7)}(1)$$

$$f^{(8)}(x) = 1680: \quad f^{(8)}(0) = f^{(8)}(1) = 1680.$$

The observations of relevance are that, in this case:

- $f^{(r)}(0) = f^{(r)}(1) = 0$ for $r < 4$.
- $f^{(r)}(0) = (-1)^r f^{(r)}(1)$ for $r \geqslant 4$.
- Both $f^{(r)}(0)$ and $f^{(r)}(1)$ are integers for all r.

To establish that this is no special case we can argue as follows.

The numerator of $f(x)$ comprises the product of two nth powers and so differentiating using the product rule fewer than n times will leave both an x term and a $(1 - x)$ term in each component of the derivative and so $f^{(j)}(0) = f^{(j)}(1) = 0$ for $0 \leqslant j < n$. Now we will deal with $f^{(j)}(x)$, where $n \leqslant j \leqslant 2n$ and do so by looking at its form in two different ways.

On the one hand, the polynomial $f(x)$ is the same as its Taylor expansion so

$$f(x) = f(0) + x f^{(1)}(0) + \frac{x^2}{2!} f^{(2)}(0) + \frac{x^3}{3!} f^{(3)}(0)$$
$$+ \frac{x^4}{4!} f^{(4)}(0) + \cdots + \frac{x^{2n-1}}{(2n-1)!} f^{(2n-1)}(0) + x^{2n}.$$

We already know that the first half of these derivatives is 0, but the coefficient of x^j is $f^{(j)}(0)/j!$ for all $0 \leqslant j \leqslant 2n$. On the other hand, if we use the standard binomial expansion on $f(x)$ we have

$$f(x) = \frac{x^n}{n!} \sum_{k=0}^{n} \binom{n}{k} 1^k (-x)^{n-k}$$
$$= \frac{1}{n!} \sum_{k=0}^{n} \binom{n}{k} (-1)^{n-k} x^{2n-k}.$$

Equating the two forms of the coefficient of x^j for $j > n$ results in

$$\frac{f^{(j)}(0)}{j!} = \frac{1}{n!} \binom{n}{2n-j} (-1)^{j-n}$$

and so

$$f^{(j)}(0) = \binom{n}{2n-j} (-1)^{j-n} \frac{j!}{n!}$$

and $f^{(j)}(0)$ is an integer.

So is $f^{(j)}(1)$. In fact, the symmetry $f(x) = f(1 - x)$ results in

$$f^{(j)}(x) = (-1)^j f^{(j)}(1 - x) \quad \text{and so} \quad f^{(j)}(1) = (-1)^j f^{(j)}(0).$$

These properties are central to Niven's approach – and that of others who followed him.

His initial use of the idea was to prove that π is irrational and, using variants of it, the irrationality of all manner of numbers fall to it. We have already demonstrated how Hermite's methods show the irrationality of π^2 but, in order to gain a feel for the use of Niven polynomials, we shall use them to re-prove the fact and then move to a further, significant application of them. Our approach follows that 1949 arguments of Y. Iwamoto.[5]

To begin with, for any differentiable function $f(x)$,

$$\int_0^1 f(x) \sin x \, dx$$

can be manipulated using integration by parts, so

$$\int_0^1 f(x) \sin \pi x \, dx$$

$$= \left[-\frac{1}{\pi} f(x) \cos \pi x \right]_0^1 + \frac{1}{\pi} \int_0^1 f^{(1)}(x) \cos \pi x \, dx$$

$$= \frac{1}{\pi} (f(0) + f(1)) + \frac{1}{\pi} \int_0^1 f^{(1)}(x) \cos \pi x \, dx.$$

Now repeat the process to get

$$\int_0^1 f(x) \sin \pi x \, dx$$

$$= \frac{1}{\pi} (f(0) + f(1)) + \frac{1}{\pi^2} [f^{(1)}(x) \sin \pi x]_0^1$$

$$- \frac{1}{\pi^2} \int_0^1 f^{(2)}(x) \sin \pi x \, dx$$

$$= \frac{1}{\pi} (f(0) + f(1)) - \frac{1}{\pi^2} \int_0^1 f^{(2)}(x) \sin \pi x \, dx,$$

---

[5]Y. Iwamoto, 1949, A proof that π^2 is irrational, *J. Osaka Inst. Sci. Tech.* 1:147–48.

which provides us with a recurrence relation, and which generates the next step,

$$\int_0^1 f(x) \sin \pi x \, dx$$

$$= \frac{1}{\pi} (f(0) + f(1))$$

$$- \frac{1}{\pi^2} \left\{ \frac{1}{\pi} (f^{(2)}(0) + f^{(2)}(1)) - \frac{1}{\pi^2} \int_0^1 f^{(4)}(x) \sin \pi x \, dx \right\}$$

$$= \left\{ \frac{1}{\pi} f(0) - \frac{1}{\pi^3} f^{(2)}(0) \right\} + \left\{ \frac{1}{\pi} f(1) - \frac{1}{\pi^3} f^{(2)}(1) \right\}$$

$$+ \frac{1}{\pi^4} \int_0^1 f^{(4)}(x) \sin \pi x \, dx$$

and the one following,

$$\int_0^1 f(x) \sin \pi x \, dx = \left\{ \frac{1}{\pi} f(0) - \frac{1}{\pi^3} f^{(2)}(0) + \frac{1}{\pi^5} f^{(4)}(0) \right\}$$

$$+ \left\{ \frac{1}{\pi} f(1) - \frac{1}{\pi^3} f^{(2)}(1) + \frac{1}{\pi^5} f^{(4)}(1) \right\}$$

$$- \frac{1}{\pi^7} \int_0^1 f^{(6)}(x) \sin \pi x \, dx$$

$$= \frac{1}{\pi^5} \left\{ \pi^4 f(0) - \pi^2 f^{(2)}(0) + f^{(4)}(0) \right\}$$

$$+ \frac{1}{\pi^5} \left\{ \pi^4 f(1) - \pi^2 f^{(2)}(1) + f^{(4)}(1) \right\}$$

$$- \frac{1}{\pi^7} \int_0^1 f^{(6)}(x) \sin \pi x \, dx$$

and continue this to the maximum $2n$th derivative to get

$$\int_0^1 f(x) \sin \pi x \, dx$$

$$= \frac{1}{\pi^{2n+1}} \left\{ \pi^{2n} f(0) - \pi^{2n-2} f^{(2)}(0) + \pi^{2n-4} f^{(4)}(0) \right.$$

$$\left. - \pi^{2n-6} f^{(6)}(0) + \cdots + (-1)^n f^{(2n)}(0) \right\}$$

$$+ \frac{1}{\pi^{2n+1}} \left\{ \pi^{2n} f(1) - \pi^{2n-2} f^{(2)}(1) + \pi^{2n-4} f^{(4)}(1) \right.$$

$$\left. - \pi^{2n-6} f^{(6)}(1) + \cdots + (-1)^n f^{(2n)}(1) \right\}$$

$$+ (-1)^{n+1} \int_0^1 f^{(2n+2)}(x) \sin \pi x \, dx.$$

Now we return to the Niven polynomial. Since $f(x)$ is a polynomial of degree $2n$, $f^{(2n+1)}(x) = f^{(2n+2)}(x) = 0$, and this means that the final integral above is 0. If we define the super-function

$$F(x) = \frac{1}{\pi^{2n+1}} \{ \pi^{2n} f(x) - \pi^{2n-2} f^{(2)}(x) + \pi^{2n-4} f^{(4)}(x)$$

$$- \pi^{2n-6} f^{(6)}(x) + \cdots + (-1)^n f^{(2n)}(x) \}$$

then

$$\int_0^1 f(x) \sin \pi x \, dx = F(0) + F(1). \tag{1}$$

Now we are ready to generate a contradiction to the assumption that $\pi^2 = a/b$. If it is so we have

$$F(x) = \frac{1}{\pi (\pi^2)^n} \{ (\pi^2)^n f(x) - (\pi^2)^{n-1} f^{(2)}(x)$$

$$+ (\pi^2)^{n-2} f^{(4)}(x) - (\pi^2)^{n-3} f^{(6)}(x)$$

$$+ \cdots + (-1)^n f^{(2n)}(x) \}$$

$$= \frac{b^n}{\pi a^n} \left\{ \frac{a^n}{b^n} f(x) - \frac{a^{n-1}}{b^{n-1}} f^{(2)}(x) + \frac{a^{n-2}}{b^{n-2}} f^{(4)}(x) \right.$$

$$\left. - \frac{a^{n-3}}{b^{n-3}} f^{(6)}(x) + \cdots + (-1)^n f^{(2n)}(x) \right\}$$

$$= \frac{1}{\pi a^n} \{ a^n f(x) - a^{n-1} b f^{(2)}(x) + a^{n-2} b^2 f^{(4)}(x)$$

$$- a^{n-3} b^3 f^{(6)}(x) + \cdots + (-1)^n f^{(2n)}(x) \}.$$

This makes

$$\pi a^n F(0) = a^n f(0) - a^{n-1} b f^{(2)}(0) + a^{n-2} b^2 f^{(4)}(0)$$

$$- a^{n-3} b^3 f^{(6)}(0) + \cdots + (-1)^n f^{(2n)}(0),$$

$$\pi a^n F(1) = a^n f(1) - a^{n-1} b f^{(2)}(1) + a^{n-2} b^2 f^{(4)}(1)$$
$$- a^{n-3} b^3 f^{(6)}(1) + \cdots + (-1)^n f^{(2n)}(1),$$

and if we recast (1) as

$$\pi a^n \int_0^1 f(x) \sin \pi x \, dx = \pi a^n F(0) + \pi a^n F(1)$$

the right-hand side is an integer and so must be the left-hand side. But consider the following.

Since $0 < x < 1$ it must be that $0 < x(1-x) < 1$ and so $0 < f(x) < 1/n!$, and this means that

$$0 < \pi a^n \int_0^1 f(x) \sin \pi x \, dx < \frac{\pi a^n}{n!} \int_0^1 \sin \pi x \, dx,$$
$$0 < \pi a^n \int_0^1 f(x) \sin \pi x \, dx < \frac{\pi a^n}{n!} \left[\frac{-\cos \pi x}{\pi} \right]_0^1 = \frac{2a^n}{n!}.$$

For n sufficiently large, this right-hand side will be less than 1 and our integer must lie between 0 and 1: a contradiction.

Now, with the philosophy that a method is a trick you use twice, Niven adapted the idea in various ways, none more important than with his proof of the irrationality of $\cos r$ for rational values of $r \neq 0$,[6] which is again to be found in the monograph. From this result there is a cascade of simple arguments (using trigonometric identities, etc.) which lead to the similar irrationality of all trigonometric functions and their inverses, the hyperbolic functions and their inverses and the exponential and natural log functions; in particular, e^r is irrational for rational $r \neq 0$ and $\ln r$ is irrational for $r \neq 1$ and $r > 0$.

He also made an intriguing observation:

> It is interesting that $\cos r$ seems to be necessarily the basic function in this process. That is, whereas the irrationality of $\sin r$ and $\tan r$ is implied at once by that of $\cos r$, it appears to be not possible in such a simple way to infer the irrationality of $\cos r$ from that of $\sin r$ or $\tan r$.

We would do well, then, to attack the matter of the irrationality of $\cos r$. Regrettably the argument, although 'elementary', is a little

---

[6]Here we take radians as the angular measure.

too long, rather too challenging and much too devious (with his choice of variant of the Niven polynomial) for it to be illuminating here, but we find an acceptable compromise in a pretty result of Alan E. Parks.[7] He too uses elementary means to extract from Niven's approach enough to establish that

If $0 < |r| \leqslant \pi$ is rational then at least one of $\sin r$ and $\cos r$ is irrational.

And he uses a variant of the Niven polynomial too; the same as Niven himself in his own proof that π is irrational

$$f(x) = \frac{x^n(a - bx)^n}{n!},$$

where a and b are positive integers.

For the moment suppose that $f(x)$ is any polynomial of degree d and that $g(x)$ is a repeatedly integrable function bounded in an interval $[0, c]$: again write

$$f^{(k)}(x) = \frac{d}{dx} f^{(k-1)}(x) \quad \text{and also} \quad g_{(k)}(x) = \int g_{(k-1)}(x)\, dx.$$

Integration by parts is set to use again with

$$\int_0^c f(x)g(x)\, dx$$

$$= [f(x)g_{(1)}(x)]_0^c - \int_0^c f^{(1)}(x)g_{(1)}(x)\, dx$$

$$= (f(c)g_{(1)}(c) - f(0)g_{(1)}(0))$$

$$\quad - \left\{ [f^{(1)}(x)g_{(2)}(x)]_0^c - \int_0^c f^{(2)}(x)g_{(2)}(x)\, dx \right\}$$

$$= (f(c)g_{(1)}(c) - f^{(1)}(c)g_{(2)}(c))$$

$$\quad - (f(0)g_{(1)}(0) - f^{(1)}(0)g_{(2)}(0))$$

$$\quad + \int_0^c f^{(2)}(x)g_{(2)}(x)\, dx,$$

---

[7] Alan E. Parks, 1986, pi, e and other irrational numbers, *American Mathematical Monthly* 93:722–23.

which continues to

$$= (f(c)g_{(1)}(c) - f^{(1)}(c)g_{(2)}(c)$$
$$+ \cdots + (-1)^d f^{(d)}(c)g_{(d+1)}(c))$$
$$- (f(0)g_{(1)}(0) - f^{(1)}(0)g_{(2)}(0)$$
$$+ \cdots + (-1)^d f^{(d)}(0)g_{(d+1)}(0))$$

with the natural stopping point of the dth derivative of $f(x)$.
  So,

$$\int_0^c f(x)g(x)\,\mathrm{d}x = F(c) - F(0),$$

where our super-function is

$$F(x) = (f(x)g_{(1)}(x) - f^{(1)}(x)g_{(2)}(x)$$
$$+ \cdots + (-1)^d f^{(d)}(x)g_{(d+1)}(x)).$$

Finally, if we reverse the stated result to its equivalent

> If $0 < |r| \leqslant \pi$ and both $\sin r$ and $\cos r$ are rational then r is irrational

we have a form we can work with.

  Suppose that $r = a/b > 0$ is any positive rational (we can use symmetry to deal with negative r) and use this numerator and denominator as the components of the Niven polynomial: let $c = r$.

  Now suppose that both $\sin r$ and $\cos r$ are rational and let the greater denominator of these two rational numbers be D, so defining the function $g(x) = D\sin x$.

  We have, then $f(x) = x^n(a - bx)^n/n!$ and $g(x) = D\sin x$, where $c = r = a/b$.

  Our previous arguments on page 118 regarding the values of $f^{(r)}(0)$ and $f^{(r)}(1)$ need little amendment to show that the equivalent is true here and so we have $f^{(k)}(0)$ and $f^{(k)}(c)$ are integral. By assumption and construction, $g^{(k)}(0)$ and $g^{(k)}(c) = g^{(k)}(r)$ are integral and again we so $F(c) - F(0)$ is an assured integer. By taking high values of n in the definition of $f(x)$ we can again make this integer lies between 0 and 1 and we have our contradiction to the rationality assumption of r.

As a welcome bonus, further prudent choices leads us to the result:

If $0 < r \neq 1$ is rational then $\ln r$ is irrational.

(Notice that, since $\ln(1/r) = -\ln r$ we may assume that $r > 1$ and so $\ln r > 0$.)

Here we let $r = a/b$ and $c = \ln r$. Define $g(x) = be^x$, then $g(0) = b$ and $g(c) = be^c = be^{\ln r} = br = b \times (a/b)a$ are integers. With the same $f(x)$ we again have an untenable situation and must conclude that $\ln r$ is irrational.

A single device has yielded fundamental results which themselves can be manipulated to provide proof of the irrational nature of very many of the basic functions of mathematics. Of course, gaps in our knowledge remain, with one such: although we now know that e^r is irrational for rational r, we know nothing about r^e.

A Return to the Surd

We will continue our search for irrational numbers with a return to surds. In chapter 1 we saw that most famous proof of the irrationality of $\sqrt{2}$, preserved by Euclid, and we hypothesized about the method of Theodorus, which may have established the irrationality of $\sqrt{3}, \sqrt{5}, \sqrt{7}, \ldots, \sqrt{15}$ and perhaps $\sqrt{17}$. What of the general case of $\sqrt{n}$? It is a small matter to extend to this by a number of methods and we will consider one which dispenses with divisibility arguments altogether and which has appeared in various forms and places throughout many years.

If $\sqrt{n}$ is an integer we are done. Otherwise, it is not and must, therefore, lie between two consecutive integers $k < \sqrt{n} < k + 1$. Now suppose that $\sqrt{n}$ is rational and that, in lowest terms, it has denominator d; d is, then, the smallest integer for which $d\sqrt{n}$ is itself an integer. Now consider $d(\sqrt{n} - k) = d\sqrt{n} - dk$, which is evidently an integer less than d; but $[d(\sqrt{n} - k)]\sqrt{n} = dn - k(d\sqrt{n})$ is an integer and we have a contradiction.

We have, then, the fundamental set of surds and we can build on them by making three easy observations:

- If $r \neq 0$ is rational and x is irrational, then rx is irrational: this ensures the irrationality of the likes of $2\sqrt{7}$ and $\frac{2}{3}\sqrt{15}$.

- If r is rational and x is irrational, then $r + x$ is irrational: this ensures the irrationality of the likes of $3 + 2\sqrt{7}$ and $\frac{4}{5} + \frac{2}{3}\sqrt{15}$.
- If r is rational then r^n is rational for all integers n; equivalently, if x is irrational, then $\sqrt[n]{x}$ is irrational for all integers n: this ensures the irrationality of the likes of $\sqrt{\sqrt{2}} = \sqrt[4]{2}$ and $\sqrt{1 + \sqrt{2}}$.

So, by combining rationals and irrationals in these ways and by root extraction, we can go some way to extending the list of known surd irrationals, but there is an obvious unanswered question: are the likes of $\sqrt{2} + \sqrt{3}$, $\sqrt{2} + \sqrt{3} + \sqrt{5}$, $\sqrt{2} + \sqrt{3} + \sqrt{5} + \sqrt{7}, \ldots$ irrational? That is, does the addition of irrational numbers necessarily result in a new irrational number?

Clearly, the answer is no; take, for example, the sum $\sqrt{2} + (2 - \sqrt{2})$. Seemingly more subtly, consider the two base 10 numbers

$$a = 0.01001000100001000001\ldots,$$
$$b = 0.10110111011110111110\ldots,$$

with the pattern continuing indefinitely. If for the time being we accept the fact (which we shall prove in chapter 8) that the decimal expansion of an irrational number is infinite and non-recurring and that for a rational number is finite or infinite and recurring, we assuredly have two irrational numbers. Now add them to get the rational number

$$a + b = 0.111111\ldots = \tfrac{1}{9}.$$

The sum of two irrational numbers is again rational. Here, the underlying structure is exposed by writing the result as

$$a + (\tfrac{1}{9} - a) = \tfrac{1}{9}.$$

in which case matters are not nearly as impressive; we have once again merely added two irrational numbers which have been constructed in a convenient manner so that the irrational part cancels out.

There is a feeling that the convenient cancellation of the irrational part of the numbers is a special device to which the 'genuine' addition of irrationals is impervious; such is the case, but we need to be more precise about things and choose to do so by following the initiative of Gregg N. Patruno,[8] who has dealt with the matter rather nicely.

If we refer back to the $\sqrt{2} + (2 - \sqrt{2})$ case above, we can remove the brackets and so consider the expression as the sum of three numbers $s = \sqrt{2} + 2 - \sqrt{2}$: the general case finds expression in the number $s = a_1 + a_2 + a_3 + \cdots + a_n$, where the $a_1^2, a_2^2, a_3^2, \ldots, a_n^2$ are rational.

Let us assume that s is rational and focus on a particular component of s, which for notational convenience we will call a_1.

Define

$$F(x, a_1, a_2, a_3, \ldots, a_n) = \prod_{\substack{\text{All combina-} \\ \text{tions of } \pm}} (x - a_1 \pm a_2 \pm a_3 \pm \ldots \pm a_n).$$

It is clear that $F(s, a_1, a_2, a_3, \ldots a_n) = 0$ but the function admits another, less obvious, property, which we begin to expose by looking at the first cases of $n = 2, 3$:

$$F(x, a_1, a_2) = \prod_{\substack{\text{All combina-} \\ \text{tions of } \pm}} (x - a_1 \pm a_2)$$

$$= (x - a_1 + a_2)(x - a_1 - a_2)$$

$$= (x - a_1)^2 - a_2^2$$

$$= (x^2 + a_1^2 - a_2^2) - a_1(2x)$$

and, if we omit the gruesome algebraic manipulation,

$$F(x, a_1, a_2, a_3) = \prod_{\substack{\text{All combina-} \\ \text{tions of } \pm}} (x - a_1 \pm a_2 \pm a_3)$$

$$= (x - a_1 + a_2 + a_3)(x - a_1 + a_2 - a_3)$$
$$\times (x - a_1 - a_2 + a_3)(x - a_1 - a_2 - a_3)$$

$$= (x^4 + 6a_1^2 x^2 - 2a_2^2 x^2 - 2a_3^2 x^2 + a_1^4$$
$$+ a_2^4 + a_3^4 - 2a_1^2 a_2^2 - 2a_1^2 a_3^2 - 2a_2^2 a_3^2)$$
$$- a_1(4a_1^2 x - 4a_2^2 x - 4a_3^2 x + 4x^3).$$

---

[8]Gregg N. Patruno, 1988, Sums of irrational square roots are irrational, *Mathematics Magazine* 61(1).

Both of which are a special case of

$$F(x, a_1, a_2, a_3, \ldots, a_n) = G(x, a_1^2, a_2^2, a_3^2, \ldots, a_n^2)$$
$$- a_1 H(x, a_1^2, a_2^2, a_3^2, \ldots, a_n^2),$$

where

$$G(x, a_1^2, a_2^2, a_3^2, \ldots, a_n^2) \quad \text{and} \quad H(x, a_1^2, a_2^2, a_3^2, \ldots, a_n^2)$$

are polynomials with integer coefficients.

The main point is that, when the right-hand side of F is multiplied out, the repeated $\pm$ ensure that the terms in $a_2, a_3, a_4, \ldots,$ a_n appear each as even powers, with only a_1 appearing to both even and odd powers; the general case is easily established using induction (or otherwise). Grouping terms accordingly provides the functions G and H, and if we evaluate at $x = s$ we get

$$F(s, a_1, a_2, a_3, \ldots, a_n)$$
$$= G(s, a_1^2, a_2^2, a_3^2, \ldots, a_n^2) - a_1 H(s, a_1^2, a_2^2, a_3^2, \ldots, a_n^2)$$
$$= 0$$

and so

$$a_1 = \frac{G(s, a_1^2, a_2^2, a_3^2, \ldots, a_n^2)}{H(s, a_1^2, a_2^2, a_3^2, \ldots, a_n^2)}.$$

Now if s is assumed to be rational we have the quotient of two rational numbers, which is itself rational, which means that a_1 is rational. Of course, the suffix 1 is arbitrary and so it must be that, if s is rational, all of the $\{a_1, a_2, a_3, \ldots, a_n\}$ are rational; put conversely, if any of the $\{a_1, a_2, a_3, \ldots, a_n\}$ is irrational then s must be irrational. So what about the example above, involving $\sqrt{2}$? The only circumstance in which the above argument can fail is when the denominator $H(s, a_1^2, a_2^2, a_3^2, \ldots, a_n^2)$ is 0 and we now identify when this can happen. Consider

$$F(s, a_1, a_2, a_3, \ldots, a_n) - F(s, -a_1, a_2, a_3, \ldots, a_n)$$
$$= 0 - F(s, -a_1, a_2, a_3, \ldots, a_n)$$
$$= [G(s, a_1^2, a_2^2, a_3^2, \ldots, a_n^2) - a_1 H(s, a_1^2, a_2^2, a_3^2, \ldots, a_n^2)]$$
$$- [G(s, a_1^2, a_2^2, a_3^2, \ldots, a_n^2) + a_1 H(s, a_1^2, a_2^2, a_3^2, \ldots, a_n^2)]$$

and so

$$-F(s, -a_1, a_2, a_3, \ldots, a_n) = -2a_1 H(s, a_1^2, a_2^2, a_3^2, \ldots, a_n^2)$$

and

$$H(s, a_1^2, a_2^2, a_3^2, \ldots, a_n^2)$$

$$= \frac{1}{2a_1} F(s, -a_1, a_2, a_3, \ldots, a_n)$$

$$= \frac{1}{2a_1} \prod_{\substack{\text{All combina-}\\ \text{tions of } \pm}} (s + a_1 \pm a_2 \pm a_3 \pm \cdots \pm a_n)$$

$$= \frac{1}{2a_1} \prod_{\substack{\text{All combina-}\\ \text{tions of } \pm}} (2a_1 + (a_2 \pm a_2)$$

$$+ (a_3 \pm a_3) + \cdots + (a_n \pm a_n))$$

$$= \frac{1}{2a_1} \prod_{T \subset \{a_2, a_3, \ldots, a_n\}} \left(2a_1 + 2 \sum_{a_r} \in T a_r\right)$$

$$= \frac{1}{a_1} \prod_{T \subset \{a_2, a_3, \ldots, a_n\}} \left(a_1 + \sum_{a_r \in T} a_r\right).$$

With this product expression for $H(s, a_1^2, a_2^2, a_3^2, \ldots, a_n^2)$ we see that it can only be 0 when some subseries $(a_1 + \sum a_r)$ vanishes – and that is just the case for our earlier examples.

Now we can add together all manner of 'natural' collections of quadratic surds to generate irrational numbers.

What of cubic surds, quartic surds, etc.? The proof of the irrationality of quadratic surds at the start of this section is easily seen to work for the general case, so that

If m and n are positive integers and if $\sqrt[m]{n}$ is not an integer then it is irrational.

But what of the sums of such surds and of mixtures of them? To be specific, what of the irrationality of the likes of $\sqrt{3} + \sqrt[3]{5}$? The water is murkier now but clears to some degree when the problem is looked at differently, by which we mean in terms of polynomial equations.

If we write $x = \sqrt{3} + \sqrt[3]{5}$ then $(x - \sqrt{3})^3 = 5$ and so $x^3 - 3\sqrt{3}x^2 + 9x - 3\sqrt{3} = 5$ and $x^3 + 9x + 5 = 3\sqrt{3}(x^2 + 1)$. Now square to get

$(x^3 + 9x + 5)^2 = 27(x^2 + 1)^2$ and we may believe the computer software when this simplifies to

$$x^6 - 9x^4 - 10x^3 + 27x^2 - 90x - 2 = 0.$$

It transpires that this polynomial equation with integer coefficients has $\sqrt{3} + \sqrt[3]{5}$ as a root and that from this fact we can discern the number's irrationality using an algebraic result known as the *rational roots theorem.*

Consider the general polynomial equation

$$a_n x^n + a_{n-1} x^{n-1} + a_{n-2} x^{n-2} + \cdots + a_0 = 0,$$

where all coefficients are integers and suppose that it has a rational root p/q, written in lowest terms. Then

$$a_n \left(\frac{p}{q}\right)^n + a_{n-1} \left(\frac{p}{q}\right)^{n-1} + a_{n-2} \left(\frac{p}{q}\right)^{n-2} + \cdots + a_0 = 0$$

and so

$$a_n p^n + a_{n-1} p^{n-1} q + a_{n-2} p^{n-2} q^2 + \cdots + a_0 q^n = 0.$$

This means that

$$a_n p^n = -q(a_{n-1} p^{n-1} + a_{n-2} p^{n-2} q + \cdots + a_0 q^{n-1})$$

and so q is a divisor of $a_n p^n$. Since p and q have no common factors, q must divide a_n.

If we rewrite the original equation as

$$a_0 q^n$$
$$= -(a_n p^n + a_{n-1} p^{n-1} q + a_{n-2} p^{n-2} q^2 + \cdots + a_1 p q^{n-1})$$
$$= -p(a_n p^{n-1} + a_{n-1} p^{n-2} q + a_{n-2} p^{n-3} q^2 + \cdots + a_1 q^{n-1})$$

we have that p is a divisor of $a_0 q^n$ and so of a_0.

In summary, the rational roots theorem states that:

> If p/q (in lowest terms) is a rational root of a polynomial equation with integer coefficients then p divides a_0 and q divides a_n.

In particular, suppose that $a_n = 1$. Then, if p/q is a rational root in lowest terms, from the above it must be that $q = \pm 1$ and so any rational root must be an integer p and one which divides the constant term a_0.

So, with our equation above, the only possible rational roots are $\pm 1, \pm 2$, none of which satisfy the equation, which ensures that $\sqrt{3} + \sqrt[3]{5}$ is irrational.

Generating that polynomial equation with integer coefficients is a variously demanding task, made a little more tractable by the thought that we only need keep track of the constant term for our purposes. Matters are easier if, rather than test the irrationality of a given number, we generate irrational numbers by writing down a polynomial equation such as

$$x^6 + 3x^5 + 2x^4 - x^3 + x^2 - 2x - 12 = 0,$$

which has two real roots ($\neq \pm 1, \pm 2, \pm 3, \pm 4, \pm 6, \pm 12$), which must therefore be irrational. Haphazard for sure, but reasonably effective too.

But let us approach matters more systematically and begin to move through the polynomials by their degree. So, with all coefficients integers, consider the following equations:

- The quadratic equation $ax^2 + bx + c = 0$. The quadratic formula provides an explicit form for possible irrational roots as $x = (-b \pm \sqrt{b^2 - 4ac})/2a$ and those which are rational must have their numerator dividing c and their denominator a.

- The cubic equation $ax^3 + bx^2 + cx + d = 0$. This is harder to deal with, but again its roots can be expressed in terms of radicals using standard formulae, one of which gives the assured real root as the unpleasant but explicit

$$x = -\frac{b}{3a} - \frac{1}{3a}\sqrt[3]{\tfrac{1}{2}\left(p + \sqrt{p^2 - 4(b^2 - 3ac)^3}\right)}$$
$$- \frac{1}{3a}\sqrt[3]{\tfrac{1}{2}\left(p - \sqrt{p^2 - 4(b^2 - 3ac)^3}\right)},$$

where $p = 2b^3 - 9abc + 27a^2 d$ and those which are rational must have their numerator dividing d and their denominator a.

- The quartic equation has its own explicit form for its roots which we dare not chronicle for fear of our typesetter's sanity but, complicated though its form is,[9] it is a combination of rationals and square and cube roots.

And here we need to take pause. We have reached the quintic equation and with it a watershed in our pursuit of irrational numbers as roots of polynomial equations, in that there is no formula for the solution of the general quintic equation. Put another way, there are quintic equations the irrational roots of which are not surds; in an important sense we have a new type of irrational number.

A New Type of Algebraic Irrational

Heralded by the inimitable Gauss in his 1799 doctoral thesis, controversially 'proved' by the obscure Paolo Ruffini (1765-1822) to various degrees of rigour between 1799 and 1813, rigorously established by the tragic Norwegian mathematician Niels Abel (1802-29) in 1824 and generalized by the equally tragic, political, love-struck and outspoken Évariste Galois (1811-32) in 1830 we have in succinct form the result:

> There are algebraic numbers which are not expressible by radicals.

Algebraic numbers are the roots of polynomial equations with integer coefficients and it is in the title of Abel's original paper[10] that matters are disclosed

> Memoir on Algebraic Equations, in which is demonstrated the impossibility of solving the general equation of the fifth degree.

This does not preclude quintics having irrational solutions which are radicals, of course. Take, for example,

$$x^5 - 5x^4 + 30x^3 - 50x^2 + 55x - 21 = 0.$$

The only possible rational solutions are integers which divide 21 and so ±1, ±3, ±7, ±21; none of which satisfy the equation. In

---

[9]The interested reader may wish to consult http://planetmath.org/encyclopedia/QuarticFormula.html for the gory details.

[10]Which appeared in *J. De Crelle*, mentioned in the previous chapter.

fact, its only real solution is the irrational

$$x = 1 + \sqrt[5]{2} - \sqrt[5]{4} + \sqrt[5]{8} - \sqrt[5]{16},$$

which is trivial compared with the innocent looking

$$x^5 - 5x + 12 = 0$$

for which the only possible rational solutions are ± 1, ± 2, ± 3, ± 4, ± 6, ± 12 and again none of them satisfy the equation, leaving the only real solution to be the irrational

$$-1.8420859661902543827118\ldots;$$

this requires about 600 symbols for its exact expression in radicals.

Yet, Abel's result has it that there are quintics (and above) for which there is no such radical expression; one example is the even more innocent looking

$$x^5 - x - 1 = 0. \qquad (*)$$

Clearly, ± 1 does not work and the equation is one subject to Abel's result; no finite expression formed of the composition of radicals will provide an explicit expression for its roots, with its only real root $\alpha = 1.1673039782614186843\ldots$.

We shall concentrate on just one of this family of irrational numbers which are algebraic non-surds, chosen for its novelty and the surprise that it exists at all.

Those readers who are acquainted with the English mathematician John Horton Conway will not be surprised to be reminded of or to learn of his off-beat *Look and Say Sequence*,[11] which is defined in the following manner:

- Choose a positive integer for the first term.
- Generate all subsequent terms of the sequence by describing the composition of the previous term.

The canonical sequence begins with 1 and so is generated as follows:

---

[11] J. H. Conway, 1985, The weird and wonderful chemistry of audioactive decay, *Eureka* 46:5–16; reprinted in 1987 in *Open Problems in Communication and Computation*, ed. T. M. Cover and B. Gopinath (Springer), pp. 173–188.

the 1st term is 1,

the 1st term is described as 'one one' and so the 2nd term is 11,

the 2nd term is described as 'two ones' and so the 3rd term is 21,

the 3rd term is described as 'one two and one one' and so the 4th term is or 1211,

the 4th term is described as 'one one, one two and two ones' and so the 5th term is or 111221,

the 5th term is described as 'three ones, two twos and one one' and so the 6th term is or 312211,

the 6th term is described as 'one three, one one, two twos and two ones' and so the 7th term is or 13112221,

the 7th term is described as 'one one, one three, two ones, three twos and one one' and so the 8th term is or 1113213211,
etc.

The sequence begins, then,

1, 11, 21, 1211, 111221, 312211, 13112221, 1113213211,...

We may suppose that the terms of the sequence become ever more complex no matter what the starting configuration, but our suspicions will be raised if we start with the sequence 22 ('two twos'), which the process leaves forever unchanged. Conway divided all possible strings of integers generated from all possible starting positions into three categories:

- 92 atomic elements, which Conway termed *common elements*;
- 2 sequences, which he termed *transuranic elements* (together with their infinite number of isotopes);
- an infinite number of *exotic elements*.

Some examples of each are:

- 22, 1113, 13112221133211322112211213322113, 132;
- 3122113222122112112322211 [≥ 4] and 1311222113321132211221121332211 [≥ 4] (the last number is any integer ≥ 4, forming the isotopes of the main element);
- 1, 2, 23, 7777.

We will sensibly demur from any attempt at a chemistry lesson but the reader with a modicum of chemical background will begin to discern a borrowing of nomenclature. The atomic (incapable of being split) nature of the 92 elements may be understood when the concept of *splitting* is defined, as below:

We will adopt a functional notation to describe the Look and Say procedure which, starting with a string of digits s, generates the string $\alpha(s)$. A string $s = lr$ splits if $\alpha(s) = \alpha(lr) = \alpha(l)\alpha(r)$ and so $\alpha^k(s) = \alpha^k(l)\alpha^k(r)$ for each positive integer k. For example,

$$\alpha(1511) = (1115)(21) = \alpha(15)\alpha(11),$$
$$\alpha^2(1511) = \alpha(111521) = (3115)(1211) = \alpha^2(15)\alpha^2(11),$$

etc.

A string which does not split is atomic – and there are precisely 92 of them. There is a Periodic Table corresponding to that of the 92 naturally occurring chemical elements, with each associated with proper atomic numbers: the table begins with hydrogen, which is associated with 22, and ends with uranium, which is associated with 3; every other member of the atomic sequence has its chemical counterpart. As to the use of *transuranic*, these strings are associated with elements with atomic numbers (93 and 94) which *transcend* that of uranium, with the borrowing of *isotope* now entirely natural. Exotic elements are the strings which are neither common nor transuranic.

Conway's self-proclaimed 'finest achievement' of the theory is his *Cosmological Theorem*, which states that, no matter the starting sequence, after a minimum of 24 iterations, the resulting sequence will consist of (be a compound of) only atomic and transuranic elements; using Conway's expression and with each iteration taking one 'day', all exotic elements will have disappeared 24 days after the 'Big Bang'.

What has all of this to do with irrational numbers? The theory generates one, as Conway proved his *Arithmetical Theorem* too, which states that the asymptotic rate of growth of the length of the sequences is constant, independent of the starting value, and that constant is

$$\lambda = 1.303577269034296\ldots$$

the *Conway Constant*. That is,

$$\frac{\#\alpha^{k+1}(s)}{\#\alpha^{k}(s)} \xrightarrow{k\to\infty} \lambda.$$

Incredibly, λ is algebraic and equally incredibly it is the single real, positive root of the polynomial equation

$$\begin{aligned}
x^{71} &- x^{69} - 2x^{68} - x^{67} + 2x^{66} + 2x^{65} + x^{64} - x^{63} - x^{62} - x^{61} \\
&- x^{60} - x^{59} + 2x^{58} + 5x^{57} + 3x^{56} - 2x^{55} - 10x^{54} - 3x^{53} \\
&- 2x^{52} + 6x^{51} + 6x^{50} + x^{49} + 9x^{48} - 3x^{47} - 7x^{46} - 8x^{45} \\
&- 8x^{44} + 10x^{43} + 6x^{42} + 8x^{41} - 5x^{40} - 12x^{39} + 7x^{38} - 7x^{37} \\
&+ 7x^{36} + x^{35} - 3x^{34} + 10x^{33} + x^{32} - 6x^{31} - 2x^{30} - 3x^{25} \\
&+ 14x^{24} - 8x^{23} - 7x^{21} + 9x^{20} + 3x^{19} - 4x^{18} - 10x^{17} - 7x^{16} \\
&+ 12x^{15} + 7x^{14} + 2x^{13} - 12x^{12} - 4x^{11} - 2x^{10} + 5x^{9} + x^{7} \\
&- 7x^{6} + 7x^{5} - 4x^{4} + 12x^{3} - 6x^{2} + 3x - 6 = 0.
\end{aligned}$$

It is also provably not expressible in surd form![12]

---

[12] Óscar Martín, 2006, Look-and-say biochemistry: exponential RNA and multistranded DNA, *American Mathematical Monthly* 113(4):289–307.

A Very Special Irrational

A proof only becomes a proof after the social act of
'accepting it as a proof'.

Yu. I. Manin

Mathematical constants are either anonymous or famous, with
fame a reflection of the constant's importance. And we can distinguish between those numbers for which fame is an intrinsic
quality and those which have had it thrust upon them: π and e
compared with $\sqrt{2}$, for example. None could doubt the star quality of the first two numbers but we have seen that it is through
Pythagorean tradition that $\sqrt{2}$ holds its distinguished place in the
mathematical firmament: this chapter is concerned with another
number whose celebrity is incidental. While we agree with the sentiment that adding two numbers that have not been added before
does not constitute a mathematical breakthrough, determining
the irrationality of this particular one assuredly did so; the result
stands as one of the most remarkable achievements of twentieth-
century mathematics. The number is written $\zeta(3)$, the second of
an infinite sequence of real numbers, defined for positive integers
n (and capable of significant extension) by

$$\zeta(n) = \sum_{r=1}^{\infty} \frac{1}{r^n}$$

with the Greek letter *zeta*, ζ, disclosing the name of the Zeta
function.

It is not, as it might appear, the third in that sequence since we owe the result of the divergence of the harmonic series

$$\zeta(1) = \sum_{r=1}^{\infty} \frac{1}{r}$$

to that medieval polymath Nicole Oresme mentioned on page 66. Nor is our interest with the first in the sequence

$$\zeta(2) = \sum_{r=1}^{\infty} \frac{1}{r^2}$$

either since in 1735 Euler had proved by quite the most ingenious means imaginable[1] that

$$\zeta(2) = \frac{\pi^2}{6}$$

and with the known irrationality of π^2 its nature is clear.

Euler achieved much more, though. He managed to prove that

$$\zeta(2n) = C\pi^{2n}$$

with C a rational constant: the irrationality of the Zeta function evaluated at positive even integers is thereby assured. What is not assured is the nature of $\zeta(2n + 1)$ for $n \geqslant 1$ since neither Euler nor any who have followed him were able to find a workable expression for the number, much less prove it to be irrational – until 1978, when the Frenchman Roger Apéry made a sequence of startling assertions which combined to a proof that

$$\zeta(3) = \sum_{r=1}^{\infty} \frac{1}{r^3} = \frac{1}{1^3} + \frac{1}{2^3} + \frac{1}{3^3} + \frac{1}{4^3} + \cdots$$

$$= 1 + \frac{1}{8} + \frac{1}{27} + \frac{1}{64} + \cdots = 1.2020569\ldots$$

is irrational, although a workable expression for the number remains elusive.

---

[1]See, for example, J. Havil, 2009, *Gamma: Exploring Euler's Constant* (Princeton University Press), p. 39.

Roger Apéry was a good mathematician. A very good mathematician. A reliable measure of quality is the opinion held of the individual by those whose reputation is already assured. That Apéry enjoyed Jean Dieudonné's friendship and the respect and the patronage of Élie Cartan is enough to make the point, should the reader wish to investigate those two gentlemen. He contributed significantly in abstract algebra, algebraic curves and Diophantine analysis, but his greatest mathematical recognition came at the age of 62, in June 1978, and began when he stood to give a lecture at the *Journées Arithmétiques* conference at Marseille-Luminy; its title, *Sur l'Irrationalité de $\zeta(3)$*:

> Though there had been earlier rumours of his claiming a proof, scepticism was general. The lecture tended to strengthen this view to rank disbelief. Those who listened casually, or who were afflicted with being non-Francophone, appeared to hear only a sequence of unlikely assertions.

So wrote Alfred van der Poorten,[2] who was one of those disbelieving listeners.

The sole purpose of this chapter is to address Apéry's original argument,[3] if not in full then nearly so, in the hope that his fantastic insights might at least be seen to combine to an implausible solution of a seemingly innocent problem; what we cannot hope to do is make his arguments natural. Once again, the proof uses contradiction: assume that $\zeta(3)$ is rational and manufacture a positive integer less than 1. That positive integer is hard-won and comes about from an extremely accurate rational approximation to $\zeta(3)$, with the discrepancy altered to an integer by multiplying it by an integer; it is this altered integer discrepancy that will prove to be eventually less than 1.

The Recurrence Relation

His argument began with a mysterious second-order recurrence relation

$$n^3 u_n + (n-1)^3 u_{n-2} = (34n^3 - 51n^2 + 27n - 5)u_{n-1}.$$

---

[2] A. van der Poorten, 1979, *Math. Intelligencer* 1:195–203.

[3] Which appeared as: R. Apéry, 1979, Irrationalité de $\zeta(2)$ et $\zeta(3)$, *Astérisque* 61:11–13.

Table 5.1.

n	a_n	b_n
0	0	1
1	6	5
2	$\dfrac{351}{4}$	73
3	$\dfrac{62531}{36}$	1445
4	$\dfrac{11424695}{288}$	33001
5	$\dfrac{35441662103}{36000}$	819005
6	$\dfrac{20637706271}{800}$	21460825
7	$\dfrac{963652602684713}{1372000}$	584307365
8	$\dfrac{43190915887542721}{2195200}$	16367912425
9	$\dfrac{150266396904385125 4939}{2667168000}$	468690849005

Two pairs of boundary conditions were associated with it, generating solutions a_n and b_n: $a_0 = 0$, $a_1 = 6$ and $b_0 = 1$, $b_1 = 5$.

Table 5.1 lists the first few terms of each of the sequences and two things are immediate: the b_n are positive integers and the a_n rational numbers, with their numerators growing quickly - and much more quickly than their denominators. To prove that this is always the case he offered, without proof, his solution to the recurrence relations:

$$b_n = \sum_{k=0}^{n} \binom{n}{k}^2 \binom{n+k}{k}^2,$$

which is easily seen to be consistent with the table. His solution for the a_n is distinctly more complex with the auxiliary sequence

$c_{n,k}$ defined by

$$c_{n,k} = \sum_{m=1}^{n} \frac{1}{m^3} + \sum_{m=1}^{k} \frac{(-1)^{m-1}}{2m^3 \binom{n}{m}\binom{n+m}{m}}$$

and

$$a_n = \sum_{k=0}^{n} c_{n,k} \binom{n}{k}^2 \binom{n+k}{k}^2.$$

We will not take the considerable trouble to prove these either, but such is the case,[4] and it becomes easy to appreciate the incredulity of those who attended that conference.

The Link to $\zeta(3)$

The expression for $c_{n,k}$ at least suggests $\zeta(3)$. In fact, his further assertion was that

$$\lim_{n \to \infty} \frac{a_n}{b_n} = \zeta(3).$$

We will look at this, first by taking a careful look at that first term of $c_{n,k}$, which is independent of the summation variable k in the expression for a_n; it can therefore be pushed through that summation, leaving it to be multiplied by the expression for b_n. Symbolically,

$$a_n = \sum_{k=0}^{n} \binom{n}{k}^2 \binom{n+k}{k}^2 \left(\sum_{m=1}^{n} \frac{1}{m^3} \right)$$

$$+ \sum_{k=0}^{n} \binom{n}{k}^2 \binom{n+k}{k}^2 \sum_{m=1}^{k} \frac{(-1)^{m-1}}{2m^3 \binom{n}{m}\binom{n+m}{m}}.$$

---

[4]The interested reader may wish to consult, for example, Jorn Steuding, 2005, *Diophantine Analysis* (Chapman & Hall), pp. 56-59.

So,

$$a_n = b_n \left(\sum_{m=1}^{n} \frac{1}{m^3} \right) + \sum_{k=0}^{n} \binom{n}{k}^2 \binom{n+k}{k}^2 \sum_{m=1}^{k} \frac{(-1)^{m-1}}{2m^3 \binom{n}{m} \binom{n+m}{m}}.$$

Now consider (for the first of two occasions) the denominator

$$2m^3 \binom{n}{m} \binom{n+m}{m}.$$

Elementary properties of Pascal's Triangle yield

$$\binom{n}{m} \binom{n+m}{m} \geqslant \binom{n}{1} \binom{n+1}{1} = n(n+1) > n^2$$

so it must be that

$$2m^3 \binom{n}{m} \binom{n+m}{m} > 2m^3 n^2$$

and so

$$\sum_{m=1}^{k} \frac{(-1)^{m-1}}{2m^3 \binom{n}{m} \binom{n+m}{m}} < \sum_{m=1}^{k} \frac{1}{2m^3 n^2}$$

$$= \frac{1}{2n^2} \sum_{m=1}^{k} \frac{1}{m^3} < \frac{1}{2n^2} \zeta(3) < \frac{1}{n^2}.$$

All of which means that

$$a_n < b_n \left(\sum_{m=1}^{n} \frac{1}{m^3} \right) + b_n \frac{1}{n^2} < b_n \zeta(3) + b_n \frac{1}{n^2},$$

which we rewrite as

$$\left| \zeta(3) - \frac{a_n}{b_n} \right| < \frac{1}{n^2}.$$

Put another way, $|\zeta(3) - a_n/b_n| \xrightarrow{n \to \infty} 0$, and we have our convergence.

The Rate of Convergence

The rate of convergence is *much* faster, though, than the above argument suggests and we shall have need of a more accurate measure of it. To that end, recall the fact that a_n and b_n each satisfy the earlier recurrence relation and so we have

$$n^3 a_n + (n-1)^3 a_{n-2} = (34n^3 - 51n^2 + 27n - 5)a_{n-1},$$
$$n^3 b_n + (n-1)^3 b_{n-2} = (34n^3 - 51n^2 + 27n - 5)b_{n-1}.$$

Now force the right-hand sides to be the same by multiplying the equations by b_{n-1} and a_{n-1} respectively to get

$$n^3 a_n b_{n-1} + (n-1)^3 a_{n-2} b_{n-1}$$
$$= (34n^3 - 51n^2 + 27n - 5)a_{n-1}b_{n-1},$$
$$n^3 a_{n-1} b_n + (n-1)^3 a_{n-1} b_{n-2}$$
$$= (34n^3 - 51n^2 + 27n - 5)a_{n-1}b_{n-1}.$$

Equating the two left-hand sides then gives

$$n^3 a_n b_{n-1} + (n-1)^3 a_{n-2} b_{n-1}$$
$$= n^3 a_{n-1} b_n + (n-1)^3 a_{n-1} b_{n-2},$$
$$n^3 a_n b_{n-1} - n^3 a_{n-1} b_n$$
$$= (n-1)^3 a_{n-1} b_{n-2} - (n-1)^3 a_{n-2} b_{n-1},$$
$$n^3 (a_n b_{n-1} - a_{n-1} b_n) = (n-1)^3 (a_{n-1} b_{n-2} - a_{n-2} b_{n-1}),$$
$$a_n b_{n-1} - a_{n-1} b_n = \frac{(n-1)^3}{n^3}(a_{n-1} b_{n-2} - a_{n-2} b_{n-1}).$$

And this recurrence relation we chase down to its end:

$$a_n b_{n-1} - a_{n-1} b_n = \frac{(n-1)^3}{n^3}\frac{(n-2)^3}{(n-1)^3}(a_{n-2} b_{n-3} - a_{n-3} b_{n-2})$$
$$= \frac{(n-2)^3}{n^3}(a_{n-2} b_{n-3} - a_{n-3} b_{n-2})$$
$$\vdots$$
$$= \frac{1}{n^3}(a_1 b_0 - a_0 b_1) = \frac{6}{n^3}.$$

Since, from the initial values, $a_1 b_0 - a_0 b_1 = 6$.

Finally, then, we have

$$a_n b_{n-1} - a_{n-1} b_n = \frac{6}{n^3},$$

which means that

$$\frac{a_n}{b_n} - \frac{a_{n-1}}{b_{n-1}} = \frac{6}{n^3 b_{n-1} b_n}.$$

Therefore,

$$\frac{a_n}{b_n} = \frac{6}{n^3 b_{n-1} b_n} + \frac{a_{n-1}}{b_{n-1}}.$$

And for later convenience we change n to $n+1$ in this expression to get

$$\frac{a_{n+1}}{b_{n+1}} = \frac{6}{(n+1)^3 b_n b_{n+1}} + \frac{a_n}{b_n}.$$

Now we form a cascading sequence

$$\frac{a_{n+1}}{b_{n+1}} - \frac{a_n}{b_n} = \frac{6}{(n+1)^3 b_n b_{n+1}},$$

$$\frac{a_{n+2}}{b_{n+2}} - \frac{a_{n+1}}{b_{n+1}} = \frac{6}{(n+2)^3 b_{n+1} b_{n+2}},$$

$$\vdots$$

$$\frac{a_N}{b_N} - \frac{a_{N-1}}{b_{N-1}} = \frac{6}{N^3 b_{N-1} b_N}.$$

Adding these gives

$$\frac{a_N}{b_N} - \frac{a_n}{b_n} = \sum_{k=n+1}^{N} \frac{6}{k^3 b_{k-1} b_k}.$$

Rewriting the earlier convergence with N replacing n, which we here leave fixed, gives

$$\lim_{N \to \infty} \frac{a_N}{b_N} = \zeta(3).$$

And so

$$\lim_{N \to \infty} \left(\frac{a_N}{b_N} - \frac{a_n}{b_n} \right) = \lim_{N \to \infty} \sum_{k=n+1}^{N} \frac{6}{k^3 b_{k-1} b_k},$$

$$\lim_{N \to \infty} \frac{a_N}{b_N} - \frac{a_n}{b_n} = \lim_{N \to \infty} \sum_{k=n+1}^{N} \frac{6}{k^3 b_{k-1} b_k},$$

$$\zeta(3) - \frac{a_n}{b_n} = \sum_{k=n+1}^{\infty} \frac{6}{k^3 b_{k-1} b_k}.$$

So,

$$\zeta(3) - \frac{a_n}{b_n} = 6 \left(\frac{1}{(n+1)^3 b_n b_{n+1}} + \frac{1}{(n+2)^3 b_{n+1} b_{n+2}} \right.$$

$$\left. + \frac{1}{(n+3)^3 b_{n+2} b_{n+3}} + \cdots \right)$$

$$\leqslant \frac{6}{b_n^2} \left(\frac{1}{(n+1)^3} + \frac{1}{(n+2)^3} + \frac{1}{(n+3)^3} + \cdots \right)$$

$$= \frac{6}{b_n^2} \left(\zeta(3) - \sum_{k=1}^{n} \frac{1}{k^3} \right)$$

$$\underset{n \to \infty}{\sim} 0 - \frac{6}{b_n^2} \zeta(3).$$

So, finally we have a much better measure for the rate of convergence:

$$\left| \zeta(3) - \frac{a_n}{b_n} \right| \sim \frac{1}{b_n^2}.$$

Such a rate will have much moment in chapter 7.

We will need to make the a_n integers, and to that end we temporarily move away from matters connected with $\zeta(3)$ to a standard and beautiful estimate of the upper bound on the size of the least common multiple of the first n positive integers.

Least Common Multiples

We will write $[1, 2, 3, \ldots, n]$ as the least common multiple of the integers $\{1, 2, 3, \ldots, n\}$.

To develop a feel for what we are about to do, let us consider a special case. Take $[1, 2, 3, \ldots, 20]$ and list the 20 component integers in factorized form as

$$2, \ 3, \ 2^2, \ 5, \ 2 \times 3, \ 7, \ 2^3, \ 3^2, \ 2 \times 5, \ 11, \ 2^2 \times 3,$$
$$13, \ 2 \times 7, \ 3 \times 5, \ 2^4, \ 17, \ 2 \times 3^2, \ 19, \ 2^2 \times 5.$$

By observation, $[1, 2, 3, \ldots, 20] = 2^4 \times 3^2 \times 5 \times 7 \times 11 \times 13 \times 17 \times 19$, the component primes of which may be broken down as

$2 \times 3 \times 5 \times 7 \times 11 \times 13 \times 17 \times 19$ All primes $\leqslant 20$,
$\qquad\quad 2 \times 3$ \qquad All primes whose squares are $\leqslant 20$,
$\qquad\qquad 2$ \qquad All primes whose cubes are $\leqslant 20$,
$\qquad\qquad 2$ \qquad All primes whose fourth powers are $\leqslant 20$.

$[1, 2, 3, \ldots, 20]$ is, we see, formed by multiplying together these four sets of numbers: primes less than 20, primes whose squares are less than 20, primes whose cubes are less than 20 and finally primes whose fourth powers are less than 20.

For a general n, using induction or otherwise, we can show that $[1, 2, 3, \ldots, n]$ is found by multiplying together all primes p, some power of which is less than or equal to n, that is,

$$[1, 2, 3, \ldots, n] = \prod_{m, p^m \leqslant n} p.$$

And what is the largest value of m for a given prime p and for a given n? We require the biggest m so that $p^m \leqslant n$. This means that $m \ln p \leqslant \ln n$ and $m \leqslant \ln n / \ln p$ and since m is to be maximal, $m = \lfloor \ln n / \ln p \rfloor$ (where we again use the Floor function).

All of this means that

$$[1, 2, 3, \ldots, n] = \prod_{p \leqslant n} p^{\lfloor \ln n / \ln p \rfloor}. \qquad (*)$$

From this we can find a measure of the size of $[1, 2, 3, \ldots, n]$.

It is evident that $p^{\lfloor \ln n / \ln p \rfloor} \leqslant n$ and so from equation $(*)$:

$$[1, 2, 3, \ldots, n] = \prod_{p \leqslant n} p^{\lfloor \ln n / \ln p \rfloor} \leqslant \prod_{p \leqslant n} n.$$

The right-hand side takes the value n to the power the number of primes less than or equal to n, with this latter number given

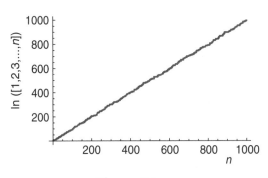

Figure 5.1.

the universally accepted symbol $\pi(n)$, with π the *prime counting function*.

If Apéry's result is one of the most significant of the twentieth century, the *Prime Number Theorem* has that distinction for the nineteenth. Conjectured by Gauss in 1792 (when he was 15 years old) it took until the end of the nineteenth century for a proof to appear; and then two such appeared simultaneously and independently in 1896: one by the Belgian mathematician Charles de la Vallée Poussin and the other by the French mathematician Jacques Hadamard.

The result takes several forms, all equivalent, and the one we need is the simplest of them:

$$\pi(n) \sim \frac{n}{\ln n}.$$

And so we have

$$[1, 2, 3, \ldots, n] \leqslant n^{\pi(n)} \sim n^{n/\ln n} = \mathrm{e}^n.$$

We now have a convenient measure of the size of $[1, 2, 3, \ldots, n]$ as n grows, with figure 5.1 graphing the trend with a plot of $\ln([1, 2, 3, \ldots, n])$.

Finally, we will need a measure of the size of the highest power of a prime p that divides a given integer N, written $\mathrm{Ord}_p(N)$. In this notation we have, then,

$$\mathrm{Ord}_p([1, 2, 3, \ldots, n]) = \left\lfloor \frac{\ln n}{\ln p} \right\rfloor$$

Table 5.2.

n	a_n	$a_n \times [1,2,3,\dots,n]^3 \times 2$
0	0	0
1	6	12
2	$\dfrac{351}{4}$	1404
3	$\dfrac{62531}{36}$	750372
4	$\dfrac{11424695}{288}$	137096340
5	$\dfrac{35441662103}{36000}$	425299945236
6	$\dfrac{20637706271}{800}$	11144361386340
7	$\dfrac{963652602684713}{1372000}$	104074481089949004
8	$\dfrac{43190915887542721}{2195200}$	23323094579273069340
9	$\dfrac{150266396904385 1254939}{2667168000}$	18031967628526215059268

and this relates to the equivalent number for binomial coefficients (which we will not prove):

$$\mathrm{Ord}_p\left(\binom{n}{m}\right) \leqslant \mathrm{Ord}_p([1,2,3,\dots,n]) - \mathrm{Ord}_p(m)$$

and so

$$\mathrm{Ord}_p\left(\binom{n}{m}\right) \leqslant \left\lfloor \frac{\ln n}{\ln p} \right\rfloor - \mathrm{Ord}_p(m).$$

Making the a_n Integers

We move closer to the contradiction that an expression is a positive integer less than 1.

To that end we shall need to address the fact that the a_n are themselves not integers but become so when multiplied by the expression $2[1,2,3,\ldots,n]^3$: this is not obvious, but as a first step we can look to table 5.2 for hope. Refer back to page 141 for the definition of a_n; $c_{n,k}$ is what makes a_n a fraction and its first component is

$$\sum_{m=1}^{n} \frac{1}{m^3} = \frac{1}{1^3} + \frac{1}{2^3} + \frac{1}{3^3} + \cdots + \frac{1}{n^3}.$$

If we were to write this as a single fraction, that fraction's denominator (before possible cancellation) would be $[1,2,3,\ldots,n]^3$ and so multiplying by this gives us an integer.

The 2 deals with the 2 in the denominator of the second part of the expression for $c_{n,k}$ but we must borrow one of the

$$\binom{n+k}{k}$$

from the expression for a_n to include with the multiplier, and so consider the multiple

$$2[1,2,3,\ldots,n]^3\binom{n+k}{k}$$

We should be clear that we are only multiplying by

$$2[1,2,3,\ldots,n]^3$$

but that the already existing

$$\binom{n+k}{k}$$

is being brought to bear in the second component of $c_{n,k}$.

Consider its essential component expression

$$2[1,2,3,\ldots,n]^3\binom{n+k}{k} \Big/ 2m^3\binom{n}{m}\binom{n+m}{m}.$$

We will need a change to the binomial element of the denominator of the expression, which is achieved by the easily checkable identity

$$\binom{n+k}{k}\Big/\binom{n+m}{m} = \binom{n+k}{k-m}\Big/\binom{k}{m},$$

which changes the expression to

$$2[1,2,3,\ldots,n]^3 \binom{n+k}{k-m} \Big/ 2m^3 \binom{n}{m}\binom{k}{m}.$$

Now we will look as the multiplicity that a given prime number appears in both the numerator and denominator of the expression. If we can show that each prime occurs to a power in the denominator which is less than that in the numerator, we shall ensure that we have an integer expression. To this end we have no need of the contribution of the

$$\binom{n+k}{k-m}$$

term. So,

$$\mathrm{Ord}_p([1,2,3,\ldots,n]) = \left\lfloor \frac{\ln n}{\ln p} \right\rfloor$$

$$\Rightarrow \mathrm{Ord}_p([1,2,3,\ldots,n]^3)$$

$$= 3\left\lfloor \frac{\ln n}{\ln p} \right\rfloor = \left\lfloor \frac{\ln n}{\ln p} \right\rfloor + \left\lfloor \frac{\ln n}{\ln p} \right\rfloor + \left\lfloor \frac{\ln n}{\ln p} \right\rfloor,$$

$$\mathrm{Ord}_p\left[m^3 \binom{n}{m}\binom{k}{m} \right]$$

$$= \mathrm{Ord}_p(m^3) + \mathrm{Ord}_p\left(\binom{n}{m}\right) + \mathrm{Ord}_p\left(\binom{k}{m}\right)$$

$$\leqslant 3\mathrm{Ord}_p(m) + \left\{ \left\lfloor \frac{\ln n}{\ln p} \right\rfloor - \mathrm{Ord}_p(m) \right\} + \left\{ \left\lfloor \frac{\ln k}{\ln p} \right\rfloor - \mathrm{Ord}_p(m) \right\}$$

$$= \mathrm{Ord}_p(m) + \left\lfloor \frac{\ln n}{\ln p} \right\rfloor + \left\lfloor \frac{\ln k}{\ln p} \right\rfloor \leqslant \left\lfloor \frac{\ln m}{\ln p} \right\rfloor + \left\lfloor \frac{\ln n}{\ln p} \right\rfloor + \left\lfloor \frac{\ln k}{\ln p} \right\rfloor.$$

Necessarily, $m \leqslant k \leqslant n$ and so each of the three components in the numerator is at least as big as each of the three components in the denominator: the prime p appears with greater multiplicity in the numerator than the denominator and so the expression must be integral and we have indeed made the a_n integers.

The Size of the b_n

We will now find an estimate for the size of the b_n for large n and to achieve this we return to the defining recurrence relation

$$n^3 b_n + (n-1)^3 b_{n-2} = (34n^3 - 51n^2 + 27n - 5)b_{n-1},$$

which we rewrite as

$$b_n + \left(\frac{n-1}{n}\right)^3 b_{n-2} = \left(34 - \frac{51}{n} + \frac{27}{n^2} - \frac{5}{n^3}\right)b_{n-1}$$

and argue that, for n large, this approaches

$$b_n + b_{n-2} = 34b_{n-1},$$

which is a nice, simple, second-order recurrence relation with constant coefficients, traditionally solved by the trial solution $b_n = B^n$ for some constant B. Substituting this into the expression results in a quadratic equation for B with $B^n + B^{n-2} = 34B^{n-1}$ or $B^2 + 1 = 34B$ or $B^2 - 34B + 1 = 0$. Using the quadratic formula it is clear that the positive solution is $B = 17 + 12\sqrt{2}$; what is not clear is that $B = 17 + 12\sqrt{2} = (1 + \sqrt{2})^4$: we have then for large n, $b_n \sim (1 + \sqrt{2})^{4n}$.

An Integer Which Isn't

Finally (!), suppose that $\zeta(3)$ is rational and in lowest terms has denominator d, and consider the expression

$$|2d[1, 2, 3, \ldots, n]^3 b_n \zeta(3) - 2d[1, 2, 3, \ldots, n]^3 a_n|,$$

which we rewrite as

$$2d[1, 2, 3, \ldots, n]^3 \times b_n \times |\zeta(3) - a_n/b_n|$$

and approximate for large n by

$$2d(e^n)^3 \times (1 + \sqrt{2})^{4n} \times \frac{1}{[(1 + \sqrt{2})^{4n}]^2}$$

$$= \frac{e^{3n}}{(1 + \sqrt{2})^{4n}} = \left(\frac{e^3}{(1 + \sqrt{2})^4}\right)^n \xrightarrow[n\to\infty]{} 0$$

since it happens to be the case that

$$e^3 = 20.08553\ldots < 33.97056\ldots = (1 + \sqrt{2})^4.$$

We have that positive integer expression which is eventually less than 1. Surely, the effect of these various estimates should be regarded but when the details are checked, all will be found to be well.

In William Blake's words:

> What is now proved was once only imagined.

In chapter 3 we gave the original proofs of the irrationality of e and π and in chapter 3 more recent, elegant and convincing arguments. Whether it is the eighteenth or the twentieth century or whatever, we have already noted that the first proof of a significant result is usually not the neatest or the shortest and it is left to those who follow to study and streamline, or to ignore altogether the original approach and replace it with their own. Is there now a proof which is more comfortable than the one we have exhibited?

Incredulous though mathematicians were, Apéry's argument was soon accepted and very quickly after that lecture the Dutch mathematician Frits Beukers produced his own proof using ideas quite different from those of Apéry,[5] and others' proofs (related to those new ideas) have followed. One such is by the Russian mathematician Wadim Zudilin and we can gain a perspective on the state of things from Zudilin's synopsis of his paper:[6]

> We present a new elementary proof of the irrationality of $\zeta(3)$ based on some recent hypergeometric ideas of Yu. Nesterenko, T. Rivoal, and K. Ball, and on Zeilberger's algorithm of creative telescoping.

At the end of six pages of well-packed symbolism representing tight arguments the author writes

> In spite of its elementary arguments, our proof of Apéry's theorem does not look simpler than the original (also elementary) Apéry's proof... or (almost elementary) Beukers's proof by means of Legendre polynomials and multiple integrals.

Mathematics makes a nice distinction between the usually synonymous terms *elementary* and *simple*, with *elementary* taken to

---

[5]F. Beukers, 1979, A note on the irrationality of $\zeta(2)$ and $\zeta(3)$, *Bull. Lond. Math. Soc.* 11:268–72.

[6]Wadim Zudilin, 2002, An elementary proof of Apéry's Theorem, E-print math.NT/0202159, 17 February.

mean that not much mathematical knowledge is needed to read the material and *simple* to mean that not much mathematical ability is needed to understand it. For this result and our purposes, the two adjectives regain their identical meaning, an antonym of which is *complex*, which for the non-specialist is precisely what these arguments remain. We will leave $\zeta(3)$ enmeshed in the mysteries of Apéry's proof and the readers to pursue the matter as fancy takes them.[7]

Alternatively, some may wish to court fame by establishing some other result connected with the irrationality of the Zeta function, but they should be aware that some progress has been made beyond Apéry, notably with the following three results:

- Infinitely many of the numbers $\zeta(3), \zeta(5), \zeta(7), \ldots$ are irrational.[8]
- Each set $\zeta(s+2), \zeta(s+4), \ldots, \zeta(8s-3), \zeta(8s-1)$ with $s > 1$ odd contains at least one irrational number.[9]
- With $s = 3$ above there is the stronger result that at least one of the four numbers $\zeta(5), \zeta(7), \zeta(9), \zeta(11)$ is irrational.[10]

We must hope that by the time that these words are read there will be more progress still.

We will leave Roger Apéry's result with the awe that it deserves and Apéry himself to rest in his tomb in the Parisian cemetery of Père Lachaise. Proximity to the tomb easily reveals its exact location, as the bottom line of the inscription the covering stone is

$$1 + \frac{1}{8} + \frac{1}{27} + \frac{1}{64} + \cdots \neq \frac{p}{q}.$$

Acquiring that proximity is quite another matter.[11]

---

[7] The above reference is a starting point, as is the following: Dirk Huylebrouck, 2001, Similarities in irrationality proofs for π, $\ln 2$, $\zeta(2)$, $\zeta(3)$, *American Mathematical Monthly* 108(3):222-31.

[8] K. Ball and T. Rivoal, 2001, Irrationalité d'une infinité de valeurs de la fonction zéta aux entiers impairs, *Invent. Math.* 146(1):193-207.

[9] W. Zudilin, 2002, Irrationality of values of the Riemann Zeta function, *Izv. Ross. Akad. Nauk Ser. Mat. (Russian Acad. Sci. Izv. Math.)* 66(3).

[10] Wadim Zudilin, 2004, Arithmetic of linear forms involving odd zeta values, *Journal de théorie des nombres de Bordeaux* 16(1):251-91.

[11] See Appendix D on page 286.

From the Rational to the Transcendental

A man is like a fraction whose numerator is what he is and whose denominator is what he thinks of himself. The larger the denominator the smaller the fraction.

Count Lev Nikolaevich Tolstoy (1828–1910)

An essential part of Apéry's proof in the previous chapter is the use of the expression $|\zeta(3) - a_n/b_n| \sim 1/b_n^2$, the left-hand side of which contains one of a sequence of rational approximations to $\zeta(3)$ and the right is a measure of the accuracy of each of those approximations. We needed only the estimate above to secure the contradiction required by his proof but in fact he had contrived a sequence of rational approximations a_n/b_n the accuracy of which is better than this and, indeed, startlingly good, as table 6.1 indicates.

How well can an irrational be approximated by a rational? If by *well* we mean *accurately*, then the answer is *as well as we like*. It is intuitively clear that the accuracy of rational approximation can, in theory, be chosen to be what we will: there are plenty of rationals and as many as we could desire as close as we desire to our chosen number; consider the decimal expansion of the irrational number, truncated as we please. Yet, there is a hidden cost, as we shall see. Alternatively, if by *well* we mean *efficiently* the story is more complex since some numbers are more amenable to rational approximation than others – and from this relative compliance we can draw important distinctions, again as we shall see. In this chapter we will look to the history, theory and practice of what is called Diophantine approximation: the study of rational approximation of irrationals, and for us this will mean positive numbers throughout. And we will give nodding recognition to the strange

Table 6.1.

n	$|\zeta(3) - a_n/b_n|$
0	1.2020569
1	2.0569×10^{-3}
2	2.10864×10^{-6}
3	1.96774×10^{-9}
4	1.77747×10^{-12}
5	1.55431×10^{-15}
6	1.396211×10^{-18}
7	1.226386×10^{-21}
8	1.073800×10^{-24}
9	9.381550×10^{-28}

idea of rational approximation of rationals too, to find them far less cooperative than the 'more awkward' irrationals. The path we will take leads to some outstanding results; that is the nature of the terrain, and the inevitable consequence of it having been trodden by some very great mathematicians of the past, several of whom will be mentioned.

Important Observations

As a preliminary, the scene is properly set by a short sequence of simple logic.

Since we are considering the rational approximation p/q of a real number α we will inevitably be interested in the error $|\alpha - p/q|$ inherent in the approximation, and if we allow a tolerance of ε we are led to the expression $|\alpha - p/q| < \varepsilon$.

For a given, fixed α, we well know that there will be an infinite number of such rational approximants but for a moment suppose the opposite is true and that there is an α for which it is otherwise. Since there are only finitely many approximants, there must be one that is closest to α: suppose it to be p_1/q_1. Certainly $|\alpha - p_1/q_1| < \varepsilon$, but now use this to define a finer tolerance $\delta = |\alpha - p_1/q_1| < \varepsilon$ and it must be that the interval $(\alpha - \delta, \alpha + \delta)$ contains no such approximant. That is, α can only be approximated up to but not exceeding a certain tolerance: we cannot approximate it as closely as we wish. For this reason we shall repeatedly be asking for there to be an infinite number of rational approximants satisfying whatever tolerance; a finite number is just not good enough to approximate with arbitrary precision.

But can this fantasy of finitely many approximants be made into reality? If we rewrite the approximation $|\alpha - p/q| < \varepsilon$ as $|\alpha q - p| < q\varepsilon$ and then use the triangle inequality we establish that

$$|p| = |(p - q\alpha) + q\alpha| \leqslant |p - q\alpha| + |q\alpha| < q\varepsilon + q\alpha = q(\varepsilon + \alpha).$$

Now consider restricting the size of the denominator of the approximating rational, say, $q \leqslant n$; then p must be bounded above also, with $p < n(\varepsilon + \alpha)$. Since we have at most n denominators and for each of them at most $\lfloor n(\varepsilon + \alpha) \rfloor$ numerators there can only be finitely many such approximants.

Put these last two facts together and we must conclude that, if the size of the denominator of the approximating rational number is restricted, the approximating process cannot be arbitrarily accurate: put the other way, if the approximation process is to be arbitrarily accurate, the denominators of the approximating rationals must increase without bound. To get arbitrarily close to α we need rationals with large denominators.

With the scene thus set we will begin our quest to ever more accurately approximate a given number, whether it be rational or irrational, by a rational number. The complexity of the approximation will naturally be judged by the size of the denominator of the approximating fraction and its accuracy measured against this: quantifying the idea we will judge that p/q is a better approximant to α than p'/q' if $q|\alpha - p/q| < q'|\alpha - p'/q'|$, that is, $|\alpha q - p| < |\alpha q' - p'|$.

Proportionate Errors

To begin with, we will ask if the accuracy of the approximation is precisely commensurate with the size of the denominator: if we are to have a complicated rational approximant, let us hope for a commensurately high accuracy in the approximation. In symbols, can we hope for an infinity of rational approximations p/q such that $|\alpha - p/q| < 1/q$ for any given real number α? Written in its alternative form, can we find integers q so that $|\alpha q - p| < 1$ for some integers p? It takes little thought to be convinced that the answer is *yes* and as a statement we have the following:

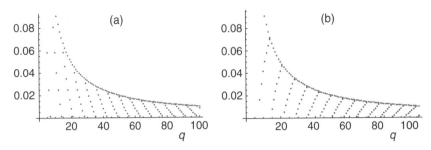

Figure 6.1.

> If α is any real number then there are infinitely many
> rationals p/q so that $|\alpha - p/q| < 1/q$.

Indeed, again using the Floor function, for any integer q we can
find a corresponding p since, if $|\alpha - p/q| < 1/q$ then $-1/q <$
$\alpha - p/q < 1/q$ and so $\alpha q - 1 < p < \alpha q + 1$, which has $p = \lfloor \alpha q \rfloor$
and $p = \lfloor \alpha q + 1 \rfloor = \lfloor \alpha q \rfloor + 1$ as integer solutions.

So, in its explicit form, we have the following:

> For any real number α and for any positive integer q the
> rational numbers $\lfloor \alpha q \rfloor / q$ and $(\lfloor \alpha q \rfloor + 1)/q$ (not necessarily
> in lowest terms) approximate α with an error less than $1/q$.

Let us break from general theory and consider an example. Suppose that fancy took us to approximate π by rational numbers
with denominators 9, 37, 113 and 1000000. Table 6.2 shows the
appropriate calculations and results and notice the outstandingly
accurate approximation of $\frac{355}{113}$.

Alternatively, parts (a) and (b) of figure 6.1 show plots of
$\pi - \lfloor \pi q \rfloor / q$ and $(\lfloor \pi q \rfloor + 1)/q - \pi$ respectively for q up to 100;
superimposed on each is the plot of $1/q$.

The Repellent Rationals

Our result makes no distinction between α being rational or irrational, so let us look at the corresponding approximants to π's
most famous rational approximant of $\frac{22}{7}$, as shown in table 6.3:
never mind any other differences, the approximation $\frac{355}{113}$ has lost
its miraculous accuracy. We must accept that there is a sharp distinction between the rational approximation of an irrational number and that of a rational approximant to it; this is inevitable since

Table 6.2.

Denominator q	Approximants	Approximating fractions	Error
9	$\dfrac{\lfloor \pi \times 9 \rfloor}{9}$ or $\dfrac{\lfloor \pi \times 9 \rfloor + 1}{9}$	$\dfrac{28}{9}$ or $\dfrac{29}{9}$	$0.030\ldots$ or $0.080\ldots$ $\left(< \frac{1}{9}\right)$
37	$\dfrac{\lfloor \pi \times 37 \rfloor}{37}$ or $\dfrac{\lfloor \pi \times 37 \rfloor + 1}{37}$	$\dfrac{116}{37}$ or $\dfrac{117}{37}$	$6.45\ldots \times 10^{-3}$ or $2.05\ldots \times 10^{-2}$ $\left(< \frac{1}{37}\right)$
113	$\dfrac{\lfloor \pi \times 113 \rfloor}{113}$ or $\dfrac{\lfloor \pi \times 113 \rfloor + 1}{113}$	$\dfrac{354}{113}$ or $\dfrac{355}{113}$	$8.8492\ldots \times 10^{-3}$ or $2.66\ldots \times 10^{-7}$ $\left(< \frac{1}{113}\right)$
1000000	$\dfrac{\lfloor \pi \times 1000000 \rfloor}{1000000}$ or $\dfrac{\lfloor \pi \times 1000000 \rfloor + 1}{1000000}$	$\dfrac{3141592}{1000000}$ or $\dfrac{3141593}{1000000}$	$6.53\ldots \times 10^{-7}$ or $3.46\ldots \times 10^{-7}$ $\left(< \frac{1}{1000000}\right)$

Table 6.3.

Denominator q	Approximants	Approximating fractions	Error
9	$\dfrac{\lfloor 22 \times 9 \rfloor}{9}$ or $\dfrac{\lfloor 22 \times 9 \rfloor + 1}{9}$	$\dfrac{28}{9}$ or $\dfrac{29}{9}$	$\dfrac{2}{63}$ or $\dfrac{5}{63}$ $\left(< \frac{1}{9}\right)$
37	$\dfrac{\lfloor 22 \times 37 \rfloor}{37}$ or $\dfrac{\lfloor 22 \times 37 \rfloor + 1}{37}$	$\dfrac{116}{37}$ or $\dfrac{117}{37}$	$\dfrac{2}{259}$ or $\dfrac{5}{259}$ $\left(< \frac{1}{37}\right)$
113	$\dfrac{\lfloor 22 \times 113 \rfloor}{113}$ or $\dfrac{\lfloor 22 \times 113 \rfloor + 1}{113}$	$\dfrac{354}{113}$ or $\dfrac{355}{113}$	$\dfrac{1}{791}$ or $\dfrac{6}{791}$ $\left(< \frac{1}{113}\right)$
1000000	$\dfrac{\lfloor 22 \times 1000000 \rfloor}{1000000}$ or $\dfrac{\lfloor 22 \times 1000000 \rfloor + 1}{1000000}$	$\dfrac{3142857}{1000000}$ or $\dfrac{3142858}{1000000}$	$\dfrac{1}{7000000}$ or $\dfrac{3}{3500000}$ $\left(< \frac{1}{1000000}\right)$

there is a sharp distinction between the rational approximation of a rational number and the rational approximation of an irrational number, as we now discuss.

Consider our standard error $|\alpha - p/q|$ but suppose that $\alpha = a/b$ is rational. Then,

$$\left| \alpha - \frac{p}{q} \right| = \left| \frac{a}{b} - \frac{p}{q} \right| = \frac{|aq - bp|}{bq}.$$

But $|aq - bp|$ is a strictly positive integer and so $|aq - bp| \geqslant 1$, which means that $|\alpha - p/q| \geqslant 1/(bq)$ and the error inherent in the approximation is bounded below; we cannot approximate with arbitrary precision. Suppose, in particular, that we wish to squeeze the proportionate error of the previous section by applying a multiple k so that $|\alpha - p/q| < 1/(kq)$; this observation tells us that it must be that $k < b$.

For example, this means that there are no rational numbers p/q ($\neq \frac{22}{7}$) which satisfy the inequality $|\frac{22}{7} - p/q| < 1/(7q)$, whereas $|\pi - p/q| < 1/(7q)$ has an infinite number of possibilities, starting with $p/q = \frac{3}{1}, \frac{22}{7}, \frac{25}{8}, \frac{47}{15}, \frac{69}{22}, \frac{113}{36}, \ldots$, ordered by size of denominator. Reducing the multiple to 6, though, opens the floodgates, with the first through them $\frac{3}{1}, \frac{25}{8}, \frac{47}{15}, \frac{69}{22}, \ldots$; that is, for example, $|\frac{22}{7} - p/q| < 1/(6q)$ is satisfied by $p/q = \frac{25}{8}$.

With $\alpha = a/b$, this lower bound on the error, $|\alpha - p/q| \geqslant 1/(bq)$, has greater implications still, since it can be rewritten as

$$\alpha - \frac{p}{q} \geqslant \frac{1}{bq} \quad \text{or} \quad \alpha - \frac{p}{q} \leqslant -\frac{1}{bq}$$

and so

$$\frac{p}{q} \leqslant \alpha - \frac{1}{bq} \quad \text{or} \quad \frac{p}{q} \geqslant \alpha + \frac{1}{bq}.$$

Now consider the implications of squeezing the error not by a multiple of q but by a power greater than 1; that is, we wish to find approximants so that

$$\left| \alpha - \frac{p}{q} \right| < \frac{1}{q^{1+\varepsilon}} \quad \text{for some } \varepsilon > 0.$$

This can be rewritten

$$\alpha - \frac{1}{q^{1+\varepsilon}} < \frac{p}{q} < \alpha + \frac{1}{q^{1+\varepsilon}}.$$

For there to be solutions to both double inequalities it must be that

$$\alpha + \frac{1}{q^{1+\varepsilon}} > \alpha + \frac{1}{bq} \quad \text{or} \quad \alpha - \frac{1}{q^{1+\varepsilon}} < \alpha - \frac{1}{bq}.$$

Both of these inequalities become $1/q^{1+\varepsilon} > 1/(bq)$, which reduces to $q^\varepsilon < b$; that is, $q < b^{1/\varepsilon}$ and so the possible values of q are $1, 2, 3, \ldots, \lfloor b^{1/\varepsilon}\rfloor$, of which there are only finitely many. And for each such q the value of p is constrained by the condition

$$\frac{|aq - bp|}{bq} < \frac{1}{q^{1+\varepsilon}} \quad \text{or} \quad |aq - bp| < \frac{b}{q^\varepsilon},$$

which simplifies to $(a/b)q - 1/q^\varepsilon < p < (a/b)q + 1/q^\varepsilon$.

So, only a finite number of values of q will suffice and to each of them there are only a finite number of values of the corresponding p; put as a mathematical statement:

> Suppose that $\alpha = a/b$ is a rational number. For any $\varepsilon > 0$ there are only finitely many rational numbers p/q such that $|\alpha - p/q| < 1/q^{1+\varepsilon}$. (1)

For example, if we return to the approximation of $\frac{22}{7}$ and take $\varepsilon = 0.9$ and so seek rationals so that $|\frac{22}{7} - p/q| < 1/q^{1.9}$ we have $q = 1, 2, 3, \ldots, \lfloor 7^{10/9}\rfloor = 1, 2, 3, 4, 5, 6, 7, 8$ and for each of these values $\frac{22}{7}q - 1/q^{0.9} < p < \frac{22}{7}q + 1/q^{0.9}$. Table 6.4 shows the calculations, where a '×' indicates that no p exists to satisfy the inequality.

And we have, perhaps a surprising, and certainly an important distinction between rational and irrational numbers: rational numbers repel rational approximation far more effectively than do irrationals. And this fact we will now pursue.

Before we do so, notice that if we are to approximate rationals to this tolerance it is never the case that αq is an integer. Were it so, with $\alpha = a/b$ the inequality becomes $|a/b - p/q| < 1/q$ and this rewritten as $|aq/b - p| < 1$ has the difference between two integers less than 1: the only possibility is that $aq/b = p$ and so $a/b = p/q$.

Table 6.4.

q	p	$\dfrac{p}{q}$	Error of $\left\lvert \dfrac{22}{7} - \dfrac{p}{q} \right\rvert$
1	3	$\dfrac{3}{1}$	$\dfrac{1}{7}\left(<\dfrac{1}{1^{1.9}}\right)$
2	6	$\dfrac{6}{2}$	$\dfrac{1}{7}\left(<\dfrac{1}{2^{1.9}}\right)$
3	$\times$	$\times$	$\times$
4	$\times$	$\times$	$\times$
5	$\times$	$\times$	$\times$
6	19	$\dfrac{19}{6}$	$\dfrac{1}{42}\left(<\dfrac{1}{6^{1.9}}\right)$
7	22	$\dfrac{22}{7}$	Not permitted
8	25	$\dfrac{25}{8}$	$\dfrac{1}{56}\left(<\dfrac{1}{8^{1.9}}\right)$

Multiple Errors

Let us temporarily return to the rational approximation of any real number: we have seen that there are infinitely many rational approximations to within an error commensurate with the size of the denominator of the approximating fraction. We will take a next step in what will be a relentless pursuit of greater accuracy of approximation by squeezing that error by a multiple and, if we content ourselves with a non-existential proof, we can halve that error very easily to yield the statement:

> If α is any real number then there are infinitely many rationals p/q so that $|\alpha - p/q| < 1/(2q)$.

This is true since, for any given real number α and positive integer q, the closed interval $[q\alpha - \frac{1}{2}, q\alpha + \frac{1}{2}]$ has length 1 and so must contain a single integer in its interior or two at its endpoints. Write p as such an integer, then $p - (q\alpha - \frac{1}{2}) < 1$ and also $(q\alpha + \frac{1}{2}) - p < 1$, so $p - q\alpha < \frac{1}{2}$ and also $p - q\alpha > -\frac{1}{2}$. Combine these to $-\frac{1}{2} < p - q\alpha < \frac{1}{2}$ and rewrite as $-1/(2q) < p/q - \alpha < 1/(2q)$ and we have our required $|\alpha - p/q| < 1/(2q)$.

Bearing in mind that natural barrier for rational approximation of rational numbers, can we increase the size of that multiple?

The answer is *yes* and to discover why has us meet the first of the great mathematicians of this chapter: Johann Peter Gustav Lejeune Dirichlet (1805–59). French before the defeat of Napoleon and Prussian after it, his mathematical pedigree is put into glittering and expert perspective by Oxford University's 11th Savilian Professor of Geometry, Henry Smith:

> The death of this eminent geometer in the present year (May 5, 1859) is an irreparable loss to the science of arithmetic. His original investigations have probably contributed more to its advancement than those of any other writer since the time of Gauss, if, at least, we estimate results rather by their importance than by their number. He has also applied himself (in several of his memoirs) to give an elementary character to arithmetical theories which, as they appear in the work of Gauss, are tedious and obscure; and he has done much to popularize the theory of numbers among mathematicians – a service which is impossible to appreciate too highly.[1]

Nothing could be more elementary than his *Schubfachprinzip* or *Drawer Principle*, which is a strong candidate for the most deceptively powerful principle in mathematics. Now widely known as the Pigeonhole Principle, its statement is:

> If n pigeons occupy m pigeonholes with $n > m$, then at least two pigeons must occupy at least one of the pigeonholes.

Any reader acquainted with the principle will need no convincing of its usefulness, and any for whom it is new has an enviably exciting and surprising journey ahead of them, should they wish to pursue the matter. As a tempter for those in the second category, consider this simple example:

> If five numbers are picked from the integers 1 to 8, then two of them must add up to nine.

To see this, form the pairs of numbers 1 and 8, 2 and 7, 3 and 6, 4 and 5; these are the four pigeon holes. The five picked numbers are the pigeons. Applying the Pigeonhole Principle, two of the five numbers must be from the same pair, which by construction sum to 9.

---

[1] H. J. S. Smith, 1965, *Report on the theory of numbers* (New York: Chelsea). (Also in *Collected papers of Henry John Stephen Smith*, vol. 1, 1894 (reprinted: New York: Chelsea, 1965).)

It seems that Dirichlet first put the principle to use in about 1834 in studying Pell's Equation[2] but we will move to 1842, when he used it to prove the first great result of Diophantine Approximation, *Dirichlet's Diophantine Approximation Theorem*:

Suppose that α is a real number, then for any positive integer n there exist positive integers p and q, with $q \leqslant n$, such that $|q\alpha - p| < 1/(n + 1)$.

The result tells us that any real number can, by multiplication by an appropriate integer, be made to approximate an integer as closely as we please. His demonstration of this took the following shape.

First we partition the semi-open interval $[0, 1)$ into $n+1$ disjoint, semi-open subintervals each of length $1/(n + 1)$ to get

$$I_1 = \left[0, \frac{1}{n + 1}\right), \quad I_2 = \left[\frac{1}{n + 1}, \frac{2}{n + 1}\right), \quad I_3 = \left[\frac{2}{n + 1}, \frac{3}{n + 1}\right),$$
$$\ldots, \quad I_n = \left[\frac{n - 1}{n + 1}, \frac{n}{n + 1}\right), \quad I_{n+1} = \left[\frac{n}{n + 1}, 1\right).$$

Now consider the set $S = [\{\alpha\}, \{2\alpha\}, \{3\alpha\}, \ldots, \{n\alpha\}]$, where $\{x\} = x - \lfloor x \rfloor$ is the fractional part of x.

First, if this set of n numbers has repetition in it there must exist i, j (with $i < j$) so that $\{i\alpha\} = \{j\alpha\}$, which means that $i\alpha - \lfloor i\alpha \rfloor = j\alpha - \lfloor j\alpha \rfloor$ and so $\alpha = (\lfloor j\alpha \rfloor - \lfloor i\alpha \rfloor)/(j - i)$ and must be rational. Evidently, $j - i < n < n + 1$ and $(j - i)\alpha - (\lfloor j\alpha \rfloor - \lfloor i\alpha \rfloor) = 0 < 1/(n + 1)$ and the condition is trivially satisfied.

We may suppose, then, that α is irrational, which means that the elements of S are distinct. The plan is to dispose of the two extreme possibilities for the location of $\{q\alpha\}$ in the interval $[0, 1)$ and then let the Pigeonhole Principle deal with the remaining alternatives.

If for some q the number $\{q\alpha\} \in I_1$ then $|\{q\alpha\}| < 1/(n + 1)$, which means that $|q\alpha - \lfloor q\alpha \rfloor| < 1/(n + 1)$, so taking $p = \lfloor q\alpha \rfloor$ gives the result.

On the other hand, if for some q the number $\{q\alpha\} \in I_{n+1}$ then $|\{q\alpha\} - 1| < 1/(n + 1)$, which means that $|q\alpha - \lfloor q\alpha \rfloor - 1| =$

[2]P. G. L. Dirichlet, R. Dedekind translation by John Stillwell: *Lectures on Number Theory* (American Mathematical Society, 1999).

$|q\alpha - (\lfloor q\alpha \rfloor + 1)| < 1/(n+1)$, so taking $p = \lfloor q\alpha \rfloor + 1$ gives the result.

Otherwise, all of the n elements of S are distributed among the $n-1$ remaining subintervals $I_2, I_3, I_4, \ldots, I_n$. Applying the Pigeonhole Principle, this means that at least two of the elements must be contained within the same subinterval: take them to be $\{q_1\alpha\}$ and $\{q_2\alpha\}$ with $q_1 < q_2$ and consider

$$|\{q_2\alpha\} - \{q_1\alpha\}| = |(q_2\alpha - \lfloor q_2\alpha \rfloor) - (q_1\alpha - \lfloor q_1\alpha \rfloor)|$$

$$= |(q_2 - q_1)\alpha - (\lfloor q_2\alpha \rfloor - \lfloor q_1\alpha \rfloor)| < \frac{1}{n+1}.$$

This time take $q = q_2 - q_1$ and $p = \lfloor q_2\alpha \rfloor - \lfloor q_1\alpha \rfloor$ and we are done. Note that $q \leqslant n$ by its construction.

Dividing throughout by q in Dirichlet's result yields:

> Suppose that α is any real number, then for any positive integer n there exist positive integers p and q, with $q \leqslant n$, such that $|\alpha - p/q| < 1/((n+1)q)$.

With this we have that the error in the approximation of α by p/q can be diminished by a multiple $n+1$ and with $q \leqslant n$.

For example, $|\pi - p/q| < 1/(101q)$ has the single solution $p = 22$, $q = 7$. But notice that the famous approximation of $\frac{22}{7}$ is, in fact, much more accurate than this, with $|\pi - \frac{22}{7}| = 1.2644\ldots \times 10^{-3}$. In fact, the result gives us much more.

From Multiples to Powers

It is that restriction on the size of q that provides a doorway between error reduction by multiples to error reduction by powers: since $q \leqslant n < n+1$ the result implies the following:

> Suppose that α is a real number, then there exist rational numbers p/q such that $|\alpha - p/q| < 1/q^2$.

So, we can demand much more of the tolerance than a mere multiple of q, we can demand its square; yet, there is no mention of our prized infinity of approximants – there cannot be, for we know it to be untrue for the rational approximation of rational numbers.

Yet, if α is irrational, suppose that there are only finitely many approximants $p_1/q_1, p_2/q_2, p_3/q_3, \ldots, p_k/q_k$ such that $|\alpha -$

Table 6.5.

q	p	$\dfrac{p}{q}$	Error of $\left| \varphi - \dfrac{p}{q} \right|$	
1	1	$\dfrac{1}{1}$	$0.618\ldots$	$\left(< \dfrac{1}{1^2} \right)$
1	2	$\dfrac{2}{1}$	$0.381\ldots$	$\left(< \dfrac{1}{1^2} \right)$
2	3	$\dfrac{3}{2}$	$0.118\ldots$	$\left(< \dfrac{1}{2^2} \right)$
3	5	$\dfrac{5}{3}$	$0.048\ldots$	$\left(< \dfrac{1}{3^2} \right)$
5	8	$\dfrac{8}{5}$	$0.018\ldots$	$\left(< \dfrac{1}{5^2} \right)$
8	13	$\dfrac{13}{8}$	$0.0069\ldots$	$\left(< \dfrac{1}{8^2} \right)$
13	21	$\dfrac{21}{13}$	$0.026\ldots$	$\left(< \dfrac{1}{13^2} \right)$

$p/q| < 1/q^2$. Since α is irrational, $|\alpha - p_r/q_r| > 0$ and so there exists a positive integer n so that $|\alpha - p_r/q_r| > 1/(n + 1)$ for all $r = 1, 2, 3, \ldots, k$. Yet Dirichlet's result asserts that, with this n, there is another rational number p/q with $q \leqslant n$ such that $|\alpha - p/q| < 1/((n + 1)q) < 1/q^2$. A contradiction.

We are led to the inevitable conclusion:

If α is irrational there are infinitely many rational numbers p/q such that $|\alpha - p/q| < 1/q^2$. $\hspace{2em}$ (2)

And here is the explicitly stated distinction between the behaviour of rational and irrational numbers: we have only finitely many rational approximants to a rational α with an error of order $q^{1+\varepsilon}$ for any $\varepsilon > 0$, yet we have an infinite number of order q^2 for any irrational α.

Draw together this result and the result which has been numbered (1) on page 160 and we have:

A real number α is irrational if and only if there are infinitely many rational numbers p/q such that $|\alpha - p/q| < 1/q^2$.

And we have a characterization of irrationality.

Table 6.6.

q	p	$\dfrac{p}{q}$	Error of $\left\lvert \sqrt{2} - \dfrac{p}{q} \right\rvert$
2	3	$\dfrac{3}{2}$	$0.085\ldots\ \left(< \dfrac{1}{2^2} \right)$
5	7	$\dfrac{7}{5}$	$0.014\ldots\ \left(< \dfrac{1}{5^2} \right)$
12	17	$\dfrac{17}{12}$	$0.0024\ldots\ \left(< \dfrac{1}{12^2} \right)$
29	41	$\dfrac{41}{29}$	$0.00042\ldots\ \left(< \dfrac{1}{29^2} \right)$
70	99	$\dfrac{99}{70}$	$0.000072\ldots\ \left(< \dfrac{1}{70^2} \right)$
169	239	$\dfrac{239}{169}$	$0.000012\ldots\ \left(< \dfrac{1}{169^2} \right)$
408	577	$\dfrac{577}{408}$	$0.0000021\ldots\ \left(< \dfrac{1}{408^2} \right)$

In table 6.5 we have the first rational approximants p/q to the Golden Ratio φ, accurate to a tolerance of $1/q^2$, and the reader will detect the appearance of the *Fibonacci Sequence*. Table 6.6 provides the same information in the approximation of $\sqrt{2}$, with a set of approximants whose pattern the reader may wish to identify.

Our pursuit of ever greater accuracy of approximation, based on the denominator of the approximating rational number, transpires to give a forensic view of our number system, with the first microscopic distinction now made: the second arises from greater demands still.

An Arithmetical Digression

For our further progress we shall need three properties of our number system.

1. A fundamental result of integer arithmetic is known as *Bézout's Identity*, which honours the eighteenth-century French mathematician Etienne Bézout (although there are several earlier contenders for primacy). In its simplest form it states that the greatest common divisor of two positive integers a and b, $\mathrm{GCD}(a, b)$, is

a linear combination of those integers; that is, there exist integers r and s so that $\text{GCD}(a, b) = br + as$.

We will not prove it (it is a direct consequence of the Euclidean Algorithm) but we do have interest in the nature of the coefficients r and s:

- First, since $br + as = (r + ka)b + (s - kb)a$ for any integer k there is an infinite number pairs of coefficients for each choice of a and b.
- Second, using the previous observation again, we can use the integer k to reduce r modulo a and s modulo b so that our choice of coefficients can be made so that $r < a$ and $s < b$.
- Third, if we choose to rewrite the result as $\text{GCD}(a, b) = br - as$ we are assured that both coefficients are positive.

We have, then, that $1 \leqslant r < a$ and $1 \leqslant s < b$.

2. The irrationals form a *dense* set, which means that between any two real numbers there is an irrational number; that is, if α and β are any two real numbers with $\alpha < \beta$, then there is an irrational number y such that $\alpha < y < \beta$. This we will prove, in part because the proof shows the deceptive power of that most subtle axiom of Eudoxus, which appears as Definition 4 in Book V of Euclid's *Elements* and which we mentioned in chapter 1. We restate it here for convenience:

> Magnitudes are said to have a ratio to one another which is capable, when a multiple of either may exceed the other.

To prove our result, suppose that $I > 1$ is any irrational number. We apply the Axiom of Eudoxus to the positive pair $\beta - \alpha$ and I. It implies that there exists a positive integer n such that $n(\beta - \alpha) > I$, or $n\beta > n\alpha + I$. Now define the positive integer m by $m = \lfloor n\alpha \rfloor$. Since $m \leqslant n\alpha$, it must be that $m + I \leqslant n\alpha + I$ and from above we have $m + I \leqslant n\alpha + I < n\beta$ and consequently $m + I < n\beta$. Since m is the greatest integer less than or equal to $n\alpha$ and also $I > 1$ it must be that $m + I > n\alpha$, which makes $n\alpha < m + I < n\beta$ and $\alpha < m/n + (1/n)I < \beta$. Since I is irrational, so must be $m/n + (1/n)I$ and the result is established.

3. The two real numbers mentioned above could themselves be rational so, given any irrational number α, we have rationals either side of it: $p/q < \alpha < r/s$ (with the fractions in lowest terms). We will now show that the fractions can be chose so that $ps - qr = 1$.

So, with α an irrational number, consider the set of rational numbers u'/v' so that $|\alpha - u'/v'| < \varepsilon$, restricting the size of the denominator $v' \leqslant n$ for n some fixed positive integer, thereby ensuring the number of them is finite (as we saw at the start of the chapter). Among this set suppose that $|\alpha - u/v|$ is minimum, with u/v in lowest terms: of course, $v \leqslant n$. There are two cases.

If $u/v < \alpha$, take u/v as the left fraction p/q and manufacture the right such as follows.

Using Bézout's Identity, there exist integers r and s with $1 \leqslant r < p$ and $1 \leqslant s < q$ such that $\mathrm{GCD}(p,q) = qr - ps$ and since $u/v = p/q$ is in lowest terms, $\mathrm{GCD}(p,q) = qr - ps = 1$. Since $qr - ps > 0$, $p/q < r/s$, and since $|\alpha - p/q|$ is minimum for $q \leqslant n$ and $s < q \leqslant n$ it must be that $|\alpha - p/q| < |\alpha - r/s|$. This means that r/s is to the right of p/q and $r/s \notin (p/q, \alpha)$ and so it must be to the right of α: this is our chosen fraction.

Otherwise $u/v > \alpha$, now take u/v to be the right fraction r/s. This time solve $qr - ps = 1$ for p and q, where $1 \leqslant p < r$ and $1 \leqslant q < s$ and the same argument reveals this p/q to be the required fraction.

It is not difficult to pursue the argument to show that there must be infinitely many such pairs.

Multiples Revisited

With the rational numbers dispensed with, it is timely to ask what we hope now is a natural question. For a given irrational number α, can we find rational approximations so that $|\alpha - p/q| < 1/kq^2$ for $k > 1$? And how many of them are there? The final result above guides us to a first answer.

With α irrational we can choose two of the infinite number of pairs of irreducible fractions so that $p/q < \alpha < r/s$, where $qr - ps = 1$.

Now define $\mu = \min\{q^2(\alpha - p/q), s^2(r/s - \alpha)\}$.

Then, by definition, $\mu \leqslant q^2(\alpha - p/q)$ and $\mu \leqslant s^2(r/s - \alpha)$ and so $\mu/q^2 \leqslant \alpha - p/q$ and $\mu/s^2 \leqslant r/s - \alpha$. Adding these results in $\mu/q^2 + \mu/s^2 \leqslant r/s - p/q$ or $\mu(1/q^2 + 1/s^2) \leqslant r/s - p/q = (qr - ps)/(qs) = 1/(qs)$. Multiply both sides by qs to get $\mu(s/q + q/s) \leqslant 1$.

We will regard this as an equation in the variable $x = s/q$ and rewrite it as $x + 1/x \leqslant 1/\mu$. But $x + 1/x \geqslant 2$ for $x > 0$ and so we have $2 \leqslant x + 1/x \leqslant 1/\mu$ and $\mu \leqslant \frac{1}{2}$.

This means that $\mu = \min\{q^2(\alpha - p/q), s^2(r/s - \alpha)\} \leqslant \frac{1}{2}$ and we have as a succinct statement:

> For every irrational number α there are infinitely many rational numbers p/q such that $|\alpha - p/q| < 1/(2q^2)$.

Tightening the Net

The tolerance has been reduced, not by a flamboyant multiple $n + 1$ of Dirichlet, but by a single multiple of 2; the terrain is becoming harder to negotiate. It was the nineteenth-century German mathematician Adolf Hurwitz (1859–1919) who navigated it, the second of the mathematical greats whom we will meet in this chapter.

A student of Felix Klein and teacher of David Hilbert and Hermann Minkowski, it is hardly surprising that Hurwitz enjoyed a high reputation as a mathematician. With over 100 research papers to his name and numerous mathematical objects and processes named after him his legacy is of the highest significance and, in number theory, succinctly encapsulated in an obituary with[3]

> His papers on continued fractions and on the approximate representation of irrational numbers are also very original, as well as curious. And all of his algebraical work is marked by a rare insight into underlying principles.

Here we will look at his 1891 result[4] which does provide a rather curious next bound for the multiple of q^2 and which we will choose to divide into two parts; the first of which pushes 2 to $\sqrt{5}$:

> For every irrational number α there are infinitely many rational numbers p/q such that $|\alpha - p/q| < 1/(\sqrt{5}q^2)$.

---

[3] *Proc. Lond. Math. Soc.* 20:xlviii–liv (1922).

[4] A. Hurwitz, 1891, Über die angenäherte Darstellung der Irrationalzahlen durch rationale Brüche, *Math. Ann.* 39:279–84.

With the above argument for the multiple of 2 in place we need do no more than modify its details to gain this extra ground.

With α irrational choose two of the infinite number of pairs of irreducible fractions so that $p/q < \alpha < r/s$, where $qr - ps = 1$. From these two fractions we manufacture a third, which is the *mediant* of them, defined as $u/v = (p + r)/(q + s)$, and which has two pertinent properties:

- it lies between the two fractions: that is,

$$\frac{p}{q} < \frac{u}{v} < \frac{r}{s} \quad \text{or} \quad \frac{p}{q} < \frac{p + r}{q + s} < \frac{r}{s};$$

- $qu - pv = 1$ and $rv - su = 1$.

The first assertion is true since

$$\frac{p + r}{q + s} - \frac{p}{q} = \frac{pq + qr - pq - ps}{q(q + s)} = \frac{qr - ps}{q(q + s)} = \frac{1}{q + s}\left(r - \frac{ps}{q}\right)$$
$$= \frac{s}{q + s}\left(\frac{r}{s} - \frac{p}{q}\right) > 0$$

(and similarly for $r/s - (p + r)/(q + s) > 0$).

And the second because

$$qu - pv = q(p + r) - p(q + s) = qr - ps = 1$$

and

$$rv - su = r(q + s) - s(p + r) = qr - ps = 1.$$

Now, since α is irrational, $u/v \neq \alpha$ and so $u/v > \alpha$ or $u/v < \alpha$. We shall deal with the possibility $u/v > \alpha$, with the alternative case nearly identical.

Define $\mu = \min\{q^2(\alpha - p/q), s^2(r/s - \alpha), v^2(u/v - \alpha)\}$. As before this means that

$$\mu \leqslant q^2\left(\alpha - \frac{p}{q}\right) \quad \text{and} \quad \mu \leqslant s^2\left(\frac{r}{s} - \alpha\right) \quad \text{and} \quad \mu \leqslant v^2\left(\frac{u}{v} - \alpha\right).$$

So

$$\frac{\mu}{q^2} \leqslant \alpha - \frac{p}{q} \quad \text{and} \quad \frac{\mu}{s^2} \leqslant \frac{r}{s} - \alpha \quad \text{and} \quad \frac{\mu}{v^2} \leqslant \frac{u}{v} - \alpha,$$

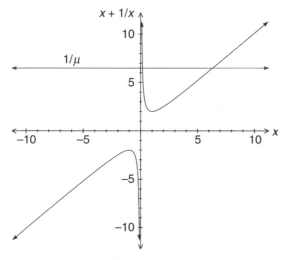

Figure 6.2.

which, added in two pairs, leads to

$$\frac{s}{q} + \frac{q}{s} \leqslant \frac{1}{\mu} \quad \text{and} \quad \frac{v}{q} + \frac{q}{v} \leqslant \frac{1}{\mu}.$$

We now return to the function $x + 1/x$, drawn in figure 6.2 together with the horizontal line at a height of $1/\mu$. If the intersections of the two occur at $x = x_1$ and $x = x_2$ (with $x_1 < x_2$) it must be that $x = s/q$ and $x = v/q$ lie in the interval $[x_1, x_2]$ and that x_1 and x_2 are the roots of the equation $x + 1/x = 1/\mu$ or $x^2 - (1/\mu)x + 1 = 0$.

Using standard elementary theory with this quadratic equation, $x_1 + x_2 = 1/\mu$ and $x_1 x_2 = 1$ and this means that $(x_2 - x_1)^2 = (x_1 + x_2)^2 - 4x_1 x_2 = 1/\mu^2 - 4$. Since $v = q + s$ and therefore $v/q = 1 + s/q$ we have $x_2 - x_1 \geqslant v/q - s/q = 1$. Put together this means that $1/\mu^2 - 4 \geqslant 1$ and so $\mu \leqslant 1/\sqrt{5}$.

All of which means that $|\alpha - p/q| < 1/(\sqrt{5}q^2)$ has an infinity of solutions for any irrational number α. At this stage we observe that the role of p can be suppressed in the result if we reformulate the inequality to $q|\alpha q - p| < 1/\sqrt{5}$ and this to $q\|\alpha q\| < 1/\sqrt{5}$, where $\|\cdots\|$ is the nearest integer function. The result then states that:

For any irrational number α, there are an infinite number of integers q for which $q\|\alpha q\| < 1/\sqrt{5}$.

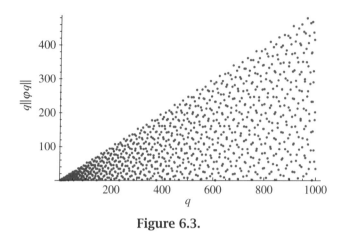

Figure 6.3.

Figure 6.3 shows a plot of $q\|\varphi q\|$, where φ is the Golden Ratio, which discloses the erratic nature of the function; nonetheless, we are guaranteed that there will be an infinity of values for q for which the plot is below $1/\sqrt{5}$. The reader may wish to check that the list of such q below 1,000,000 is

$$\{1, 3, 8, 21, 55, 144, 377, 987, 2584, 6765,$$
$$17711, 46368, 121393, 317811, 832040\}.$$

Finally, with this reformulation, for those who are comfortable with the mathematics, we are able to present one of the most famous conjectures in the whole of the theory of Diophantine approximation.

We are now well aware that for any irrational number α there are infinitely many integers q for which $q\|\alpha q\| < 1$: this clearly implies that, for any two irrational numbers α and β, there are infinitely many integers q for which $q\|\alpha q\|\,\|\beta q\| < 1$. Figure 6.4 displays the near mesmeric behaviour of the function of q where $\alpha = \varphi$ and $\beta = \pi$ and hardly suggests what, in the 1930s, John Edensor Littlewood conjectured: $\liminf_{q\to\infty} q\|\alpha q\|\,\|\beta q\| = 0$. Fame awaits.

The Irrational Cascade

So far this process of rational approximation has distinguished between rational and irrationals; with the second part of Hurwitz's

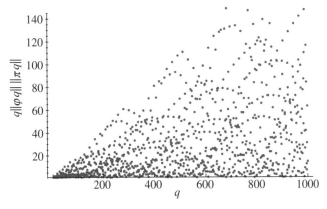

Figure 6.4.

result we begin to distinguish between the irrationals themselves. In words, the inequality is tight; in symbols:

> If $c > \sqrt{5}$ there exist irrational numbers α for which $|\alpha - p/q| < 1/(cq^2)$ has only finitely many possibilities for p/q.

In fact, one such irrational is the Golden Ratio $\alpha = \varphi = (1+\sqrt{5})/2$.

To see this, suppose that $|(1 + \sqrt{5})/2 - p/q| < 1/(cq^2)$ has in infinite number of solutions for p and q for some constant $c > \sqrt{5}$ and define the variable x as a function of p and q by $(1 + \sqrt{5})/2 - p/q = 1/(xq^2)$. It must be, then, that $|x| > c > \sqrt{5}$. Rewrite the equation as $1/(xq) - \sqrt{5}q/2 = q/2 - p$ and square to give $(1/(xq) - \sqrt{5}q/2)^2 = (q/2 - p)^2$ and so $1/(x^2q^2) - \sqrt{5}/x + 5q^2/4 = q^2/4 - pq + p^2$, to arrive at $p^2 - pq - q^2 = 1/(x^2q^2) - \sqrt{5}/x$ and $|p^2 - pq - q^2| = |1/(x^2q^2) - \sqrt{5}/x|$.

The left-hand side of the equation is an integer and so must be the right-hand side, but

$$\left| \frac{1}{x^2q^2} - \frac{\sqrt{5}}{x} \right| \leqslant \left| \frac{1}{x^2q^2} \right| + \left| \frac{\sqrt{5}}{x} \right| < \frac{1}{q^2} + \frac{\sqrt{5}}{c}$$

and since $\sqrt{5}/c < 1$ we can choose from the infinite number of possibilities for q one so large that the whole right-hand side of the inequality is less than 1, which, for such q, makes $|p^2 - pq - q^2|$ an integer less than 1 and so $p^2 - pq - q^2 = 0$ and $(p/q)^2 - p/q - 1 = 0$, the defining equation for the Golden Ratio itself; $p/q = (1 + \sqrt{5})/2 = \varphi$. Our 'rational' approximation has to be irrational and we are done.

With this we classify the Golden Ratio as the first 'awkward' irrational, that is, the first that does not allow increasing the multiple of q^2 beyond $\sqrt{5}$. Is it alone in this? Does omitting it, and perhaps others as awkward, allow us to decrease the tolerance by increasing the multiple of q^2 at will? The answer to both questions is *no*; there are an infinite number of numbers as awkward as the Golden Ratio and an infinite number more sharing an infinite number of lesser degrees of awkwardness. It is all a very complicated story, really. To see just how complicated, we start with the list of multiples in the denominator, which has been given the name the Lagrange Spectrum,

$$\{\sqrt{5}, \sqrt{8}, \tfrac{\sqrt{221}}{5}, \tfrac{\sqrt{1517}}{13}, \dots\}$$

and match it to the corresponding list of awkward numbers

$$\{\tfrac{-1+\sqrt{5}}{2}, -1+\sqrt{2}, \tfrac{-9+\sqrt{221}}{14}, \tfrac{-23+\sqrt{1517}}{38}, \dots\}.$$

With these we know that:

- For every irrational number α other than φ there are infinitely many rational numbers p/q such that $|\alpha - p/q| < 1/(cq^2)$, where $c > \sqrt{5}$.
- For every irrational number α other than φ and $-1+\sqrt{2}$ there are infinitely many rational numbers p/q such that $|\alpha - p/q| < 1/(cq^2)$, where $c > \sqrt{8}$.
- For every irrational number α other than φ, $-1+\sqrt{2}$ and $(-9+\sqrt{221})/14$ there are infinitely many rational numbers p/q such that $|\alpha - p/q| < 1/(cq^2)$ where $c > \sqrt{221}/5$.
- For every irrational number α other than φ, $-1+\sqrt{2}$, $(-9+\sqrt{221})/14$ and $(-23+\sqrt{1517})/38$ there are infinitely many rational numbers p/q such that $|\alpha - p/q| < 1/(cq^2)$ where $c > \sqrt{1517}/13$, etc.

Apart from noticing that there are plenty of square roots and quadratic irrationals around, we suggest that the pattern is opaque; we also suggest that it is barely less so when we reveal that pattern. For this purpose it is appropriate to divide the real line into four intervals, as below.

$\{x: x < \sqrt{5}\}$ This contains no element of the spectrum.

{x: $\sqrt{5} \leqslant x < 3$} Here there are a countably infinite number of elements of the spectrum, each given by the following prescription.

The general term of the infinite sequence is $(\sqrt{9w^2 - 4})/w$ with each w defined to be the maximum of successive triplets $\{u, v, w\}$ of positive integer solutions of the Diophantine equation

$$u^2 + v^2 + w^2 = 3uvw.$$

It is hardly obvious but the sequence of solutions to the equation (in order) is

$$(u, v, w) \in \{(1, 1, 1), (1, 1, 2), (1, 2, 5), (1, 5, 13), (2, 5, 29), \dots\}$$

making the *Markov numbers w*

$$\{1, 2, 5, 13, 29, 34, \dots\}.$$

Substitute them into the formula and we see that previous sequence of tolerances generated.

{x: $3 \leqslant x < \theta$} This is the region which is a subject of current research, some of the fruits of which are

- $\theta = 4 + \frac{253589820 + 283748\sqrt{462}}{491993569}$, proved by G. Freiman,[5] and therefore appropriately known as *Freiman's constant*.
- If α is not a quadratic irrational then the corresponding $c \geqslant 3$.
- There are uncountably infinite α with $c = 3$.
- Oskar Perron proved that $(\sqrt{12}, \sqrt{13})$ and $(\sqrt{13}, \frac{65+9\sqrt{3}}{22})$ are maximal *gaps*, that is, there is no α for which $c \in (\sqrt{12}, \sqrt{13})$ or $c \in (\sqrt{13}, \frac{65+9\sqrt{3}}{22})$. (There are other such too, for example, $(\sqrt{\frac{480}{7}}, \sqrt{10})$).
- $c = \sqrt{13}$ precisely when $\alpha = \frac{-1+\sqrt{13}}{2}$ (or its equivalent), yet there are uncountably many non-equivalent α for which $c = \sqrt{12}$.

---

[5] G. Freiman, 1975, *Diophantine approximations and the geometry of numbers (Markov's problem)* (Kalinin Gosudarstv. University).

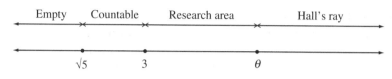

Figure 6.5.

$\{x: x \geqslant \theta\}$ The whole half line is in the spectrum. The region is called *Hall's Ray*, after Marshall Hall,[6] who proved it to be so. Figure 6.5 displays the situation.

Finally, what is this concept of equivalence to which we have alluded? It is encapsulated in the following definition:

> Two real numbers α and β are said to be *equivalent* if there exist integers p, q, r, s with $|ps - qr| = 1$ so that $\beta = (p\alpha + q)/(r\alpha + s)$.

The opacity is no less, but this does define an equivalence relation[7] in the technical sense and so divides the reals into an infinite number of disjoint equivalence classes, one of which is the set of rational numbers, and any two equivalent numbers are capable of the same degree of rational approximation.

Quadratic Irrationals

We are wallowing in deep mathematical water, where statements are barely believable and proofs barely understandable. We can gain a firmer hold on matters if we move nearer to the shallows and see the shape of these ideas when they are applied to a special kind of irrational number: the quadratic irrational; that is, an irrational number which is the root of a quadratic equation with integer coefficients.

So, let α be a quadratic irrational and suppose that its defining equation is $ax^2 + bx + c = 0$ and that the discriminant of this quadratic in x is $\Delta_x = b^2 - 4ac$.

First, with α irrational and the coefficients of the quadratic expression integers there are restrictions on Δ_x. It is clear that it cannot be a perfect square. Nor can it be of the form $4n + 2$. Were it so, it would mean that b^2 is even and so $b = 2m$ must be

---

[6]Marshall Hall Jr, 1947, On the sum and product of continued fractions, *Ann. of Math.* (2) 48:966–93.

[7]See Appendix E on page 289.

even and this makes $4n + 2 = 4m^2 - 4ac$ or $1 = 2m^2 - 2ac - 2n$, which is evidently impossible in integers; similarly it cannot be of the form $4n + 3$. There is no such contradiction for $\Delta_x = 4n$ or $\Delta_x = 4n + 1$ and so the possibilities for the discriminant are the positive integers of the form $4n$ or $4n + 1$ and which are not perfect squares:

$$\{5, 8, 12, 13, 17, 20, 21, \ldots\}.$$

Now we will show that the discriminant of the quadratic equation in the transformed variable $w = (px + q)/(rx + s)$ is again Δ_x.

It is simple to check that the composite (bilinear) transformation can be decomposed into three simpler transformations, in this order:

$$x \to rx + s, \quad x \to \frac{1}{x}, \quad x \to \frac{p}{r} - \frac{ps - qr}{r}x.$$

With the extra condition $|ps - qr| = 1$ we arrive at

$$x \to rx + s, \quad x \to \frac{1}{x}, \quad x \to \frac{p}{r} \pm \frac{1}{r}x.$$

Now consider what effect each transformation has individually, and so in combination, on the discriminant of the original quadratic equation:

With the first transformation we can generate the transformed quadratic equation from the original by writing $y = rx + s$ and so $x = (y - s)/r$ and substituting to get

$$a\left(\frac{y - s}{r}\right)^2 + b\left(\frac{y - s}{r}\right) + c = 0$$

or $a(y - s)^2 + br(y - s) + cr^2 = 0$ and finally

$$ay^2 + (br - 2as)y + (as^2 - brs + cr^2) = 0,$$

which is another quadratic equation and has discriminant

$$\begin{aligned}\Delta_y &= (br - 2as)^2 - 4a(as^2 - brs + cr^2)\\ &= b^2r^2 - 4abrs + 4a^2s^2 - 4a^2s^2 + 4abrs - 4acr^2\\ &= b^2r^2 - 4acr^2 = r^2(b^2 - 4ac) = r^2\Delta_x.\end{aligned}$$

The transformation $z = 1/y$ and so $y = 1/z$ of the quadratic equation yields another quadratic equation

$$a\left(\frac{1}{z}\right)^2 + (br - 2as)\frac{1}{z} + (as^2 - brs + cr^2) = 0$$

or $(as^2 - brs + cr^2)z^2 + (br - 2as)z + a = 0$ with integer coefficients and it is easy to see that its discriminant $\Delta_z = \Delta_y$.

The final transformation of this equation $w = p/r \pm (1/r)z$ is just a repeat of the first and yields a final quadratic equation with integer coefficients with discriminant $\Delta_w = (1/r^2)\Delta_z$. The discriminant of this final quadratic equation, obtained from the original under the combined transformations is then

$$\Delta_w = \frac{1}{r^2}\Delta_z = \frac{1}{r^2}\Delta_y = \frac{1}{r^2}r^2\Delta_x = \Delta_x.$$

And what relevance has this nice fact to us?

Rewrite the defining quadratic function for α as $f(x) = ax^2 + bx + c = a(x - \alpha)(x - \beta)$, where β is the other (irrational) root.

Now suppose that p/q is a rational approximant to α such that $|\alpha - p/q| < 1/(cq^2)$, then, using the triangle inequality,

$$\left|f\left(\frac{p}{q}\right)\right| = \left|a\left(\frac{p}{q} - \alpha\right)\left(\frac{p}{q} - \beta\right)\right| = \left|\left(\frac{p}{q} - \alpha\right)\right|\left|a\left(\frac{p}{q} - \beta\right)\right|$$

$$= \left|\alpha - \frac{p}{q}\right|\left|a\left(\beta - \frac{p}{q}\right)\right|$$

$$= \left|\alpha - \frac{p}{q}\right|\left|a\left(\beta - \alpha + \alpha - \frac{p}{q}\right)\right|$$

$$= \left|\alpha - \frac{p}{q}\right|\left|a\left(\beta - \alpha\right) + a\left(\alpha - \frac{p}{q}\right)\right|$$

$$< \left|\alpha - \frac{p}{q}\right|\left\{\left|a\left(\beta - \alpha\right)\right| + \left|a\left(\alpha - \frac{p}{q}\right)\right|\right\}$$

$$= \left|\alpha - \frac{p}{q}\right|\left\{\left|a\left(\alpha - \beta\right)\right| + \left|a\left(\alpha - \frac{p}{q}\right)\right|\right\}.$$

We can use the standard relations $\alpha + \beta = -b/a$ and $\alpha\beta = c/a$ to rewrite the discriminant in terms of the two roots as follows:

$$\Delta = b^2 - 4ac = a^2\left(\frac{b^2}{a^2} - 4\frac{c}{a}\right) = a^2((\alpha + \beta)^2 - 4\alpha\beta) = a^2(\alpha - \beta)^2.$$

And this means that

$$\left| f\left(\frac{p}{q}\right) \right| < \frac{1}{cq^2}\left(\sqrt{\Delta} + \frac{|a|}{cq^2}\right).$$

But also

$$\left| f\left(\frac{p}{q}\right) \right| = \left| a\left(\frac{p}{q}\right)^2 + b\left(\frac{p}{q}\right) + c \right| = \left| \frac{ap^2 + bpq + cq^2}{q^2} \right|$$

$$= \frac{|ap^2 + bpq + cq^2|}{q^2} \geq \frac{1}{q^2}$$

since the numerator must be an integer.

Combining these two inequalities for $|f(p/q)|$ results in

$$\frac{1}{cq^2}\left(\sqrt{\Delta} + \frac{|a|}{cq^2}\right) > \frac{1}{q^2}$$

and so $\sqrt{\Delta} + |a|/(cq^2) > c$ and finally $cq^2(c - \sqrt{\Delta}) < |a|$.

Now, if $c > \sqrt{\Delta}$, as q increases without bound the left side of the inequality is unbounded, yet the right has the fixed value of $|a|$, which is plainly impossible. The only possible reconciliation is that q is bounded above, leaving only finitely many possibilities for the denominator of the approximating fraction and accordingly only finitely many for the numerator, as we have seen earlier.

In summary:

> If α is a quadratic irrational and the associated quadratic equation has discriminant Δ then there are only finitely many rationals satisfying $|\alpha - p/q| < 1/(cq^2)$ when $c > \sqrt{\Delta}$.

We have located some of those elements of the Lagrange Spectrum. The $\sqrt{5}$ and $\sqrt{8}$ are in the first non-empty interval, the $\sqrt{12}$, $\sqrt{13}$, $\sqrt{17}$ and $\sqrt{20}$ in the next and the remainder in the Hall Ray.

Of course, we can use these discriminants to generate the corresponding 'awkward' irrational with $\Delta = b^2 - 4ac = 5$ and so it must be that b is odd. Suppose that $b = 1$, then $ac = -1$ and we have the two possible quadratic equations $x^2 - x - 1 = 0$ and $x^2 + x - 1 = 0$, the first equation yielding its positive root as the Golden Ratio, $\varphi = (1 + \sqrt{5})/2$ and the second its reciprocal

$$\frac{1}{\varphi} = \frac{-1 + \sqrt{5}}{2} = \frac{\varphi - 1}{0\varphi + 1},$$

equivalent (as we defined earlier) to φ: other odd values of b yield other, equivalent, awkward irrationals.

As to the case $c = \sqrt{13}$ mentioned in the list on page 175: take $\Delta = 13$ and with $b^2 - 4ac = 13$ again it must be that b is odd and we take $b = 1$ to yield $ac = -3$ and choosing $a = 1, c = -3$ yields the quadratic equation $x^2 + x - 3 = 0$ with positive root $(-1 + \sqrt{13})/2$: again, all other possibilities are equivalent to this.

The Finest of Filters

The move that increases the multiple of q^2 from 2 to $\sqrt{5}$ and then beyond has brought with it great consequences, embodied in that fantastically complicated Lagrange Spectrum of critical multiples. We now take what is to be a final step in our refinement of the tolerance of rational approximation by increasing the power beyond 2 and asking, for a given irrational number α, whether, for some $\varepsilon > 0$, there are an infinite number of rational approximations so that

$$|\alpha - p/q| < 1/q^{2+\varepsilon}.$$

In fact, we have done enough to show that this added requirement is enough to eliminate quadratic irrationals.

Recall that for a quadratic irrationals there are only finitely many rationals satisfying $|\alpha - p/q| < 1/(cq^2)$ when $c > \sqrt{\Delta}$. This means that we can increase c to a number c_1 so that there are no rational numbers for which $|\alpha - p/q| < 1/(c_1 q^2)$ and from this we can conclude that:

> If α is a quadratic irrational then there are only finitely many rational numbers for which $|\alpha - p/q| < 1/q^{2+\varepsilon}$ for any $\varepsilon > 0$.

To see this, suppose otherwise and that there are infinitely many such p/q. This means that q will increase without bound and we can choose a $q = q_1$ for which $q_1^\varepsilon > c_1$. But then

$$\left| \alpha - \frac{p_1}{q_1} \right| < \frac{1}{q_1^{2+\varepsilon}} = \frac{1}{q_1^\varepsilon q_1^2} < \frac{1}{c_1 q_1^2}.$$

We have our contradiction.

With this impetus, what of cubic irrationals, quartic irrationals, and in general *algebraic irrationals* of degree n: irrational numbers which are the roots of an irreducible polynomial of degree

n having integer coefficients? With the preparation that we have taken, the quadratic irrationals have fallen easily but no amount of careful structuring will account in any way easily for the algebraics in general. The original proof of the result covers 20 journal pages and is of the greatest depth and subtlety, with modern variants of it hardly less transparent or inviting; its statement:

> For any algebraic number α of degree $\geqslant 2$ and for any $\varepsilon > 0$ there are only finitely many rational numbers p/q such that $|\alpha - p/q| < 1/q^{2+\varepsilon}$.

What we have proved for quadratic irrationals is true for algebraic irrationals in general. It was the German Klaus Roth who first proved the result in 1955[8] and for it he was awarded the Fields Medal, the mathematical equivalent of the Nobel Prize. The significant British number theorist Harold Davenport spoke at the presentation of the award at the International Congress in Edinburgh in 1958 and his words sum up the level of the achievement rather succinctly:

> The achievement is one that speaks for itself: it closes a chapter, and a new chapter is now opened. Roth's theorem settles a question which is both of a fundamental nature and of extreme difficulty. It will stand as a landmark in mathematics for as long as mathematics is cultivated.

That new chapter is now well-read, if far from understood, and is host to many results and more conjectures – most of which we shall leave untouched. What remains though after this final, finest sieving are the irrationals which are not the roots of algebraic equations: the transcendentals.

---

[8]K. F. Roth, 1955, Rational approximations to algebraic numbers, *Mathematika* 2:1–20.

Transcendentals

If equations are trains threading the landscape of numbers,
then no train stops at pi.

Richard Preston

In the last chapter we located the country of the transcendentals
without identifying any of its inhabitants: here we will discover
all manner of transcendental numbers from the lowly no-names
to the important such as π and e.

The concept of transcendence appears to date back to Leibniz
since, in 1682, he discussed $\sin x$ not being an algebraic function
of x (that is, it cannot be written as a finite composition of powers,
multiples, sums and roots of x) and in 1704 he stated that

the number $\alpha = 2^{1/a}$, $a = \sqrt{2}$ is intercendental

without, however, defining the term.

And recall the comment of John Wallis on page 91. Euler's monumental work of 1744 *Introduction to the Analysis of the Infinite*
has in it:

Since the logarithms of (rational) numbers which are not powers of the base are neither rational nor irrational,[1] it is with
justice that they are called transcendental quantities. For this
reason logarithms are said to be transcendental.

and

For instance, if c denoted the circumference of a circle of radius
equal to 1, c will be a transcendental quantity.

---

[1] Meaning *algebraic.*

There are very many other references to this most elusive concept from these and from other mathematicians of the seventeenth and eighteenth centuries: much was suspected but nothing proved.

The First Transcendental

Not until 1844 and, more conveniently stated, 1851 was a provably transcendental number exhibited and it was not π, e or any other illustrious constant but an entirely contrived number: a *Liouville Number*.

That $d^{1/2}/dx^{1/2}(x) = 2\sqrt{x}/\sqrt{\pi}$ is a startling fact suggested by Leibniz in a letter to L'Hôpital dated 1695, yet the first rigorous justification of this result of the *fractional calculus* was provided by Joseph Liouville (1809–82) in 1832 and, in keeping with his eclectic scientific interests, included in that memoir problems in mechanics which yielded to an approach using fractional derivatives. He is also remembered for the disquieting fact that combinations of elementary functions do not necessarily possess elementary anti-derivatives; that is, integration is intrinsically hard. Liouville was at once physicist, astronomer, pure mathematician, founder and long time editor of a greatly influential mathematics magazine[2] and politician. And he was the first person to provide that thing so long suspected: a transcendental number.

In 1844, using ideas from the theory of continued fractions, Liouville had shown how to construct transcendental numbers without explicitly exhibiting any one of them,[3] but in 1851 he finally was able to produce a number which was provably transcendental[4] and did so using what is now known as the *Liouville Approximation Theorem*. Roth's result, of the previous chapter, was the culmination of a sequence of refinements to the result, the statement of which is:

---

[2] *Journal de Mathématiques Pures et Appliquées.*

[3] J. Liouville, 1844, Sur les classes très étendues de quantités dont la valeur n'est ni algébrique ni même réductible à des irrationelles algébriques, *Comptes Rendus Acad. Sci. Paris* 18:883–85, 910–11.

[4] J. Liouville, 1851, Sur des classes très-étendues de quantités dont la valeur n'est ni algébrique, ni même réductible à des irrationelles algébriques, *J. Math. Pures Appl.* 16:133–42.

If α is an irrational root of an irreducible polynomial

$$f(x) = a_n x^n + a_{n-1}x^{n-1} + a_{n-2}x^{n-2} + \cdots + a_0$$

with integer coefficients, then for any $\varepsilon > 0$ there are only finitely many rational numbers p/q so that $|\alpha - p/q| < 1/q^{n+\varepsilon}$.

We see that it is a much weaker form of Roth's result, with the tolerance dependent on the degree of the algebraic irrational's defining polynomial. Liouville himself suspected the result was not optimal in that the power of q may well be capable of reduction. Reduced it was, stage by stage, with:

- Axel Thue (1909) reduced n to $\frac{1}{2}n + 1$;
- Carl Siegel (1921) reduced $\frac{1}{2}n + 1$ to $2\sqrt{n}$;
- Freeman Dyson and Aleksandr Gelfond (independently in 1947) reduced $2\sqrt{n}$ to $\sqrt{2n}$.

And then came that ultimate result of Roth's.

Yet, comparatively weak though the Liouville result is, it provided a *necessary* condition for a number to be algebraic, and therefore a *sufficient* condition for a number to be transcendental and, since we are in the novel position of being able to prove a general result, this we will now do.

Referring to the above statement, choose an interval $(\alpha - \delta, \alpha + \delta)$ around α small enough to ensure that α is the only root of $f(x)$ in that interval: if p/q is a rational approximation to α either it lies in this interval or it does not.

Suppose that it does not, then $\delta < |\alpha - p/q| < 1/q^{n+\varepsilon}$ and this means that $q < 1/\delta^{1/(n+\varepsilon)}$.

Otherwise, p/q does lie in the interval and

$$\left| f\left(\frac{p}{q}\right) \right| = \left| a_n \left(\frac{p}{q}\right)^n + a_{n-1}\left(\frac{p}{q}\right)^{n-1} + a_{n-2}\left(\frac{p}{q}\right)^{n-2} + \cdots + a_0 \right|$$

$$= \left| \frac{a_n p^n + a_{n-1}p^{n-1}q + a_{n-2}p^{n-2}q^2 + \cdots + a_0 q^n}{q^n} \right|,$$

which means that $|f(p/q)| \geqslant 1/q^n$, yet the Mean Value Theorem[5] tells us that there is a number c between α and p/q so that $f(p/q) - f(\alpha) = (p/q - \alpha)f'(c)$. This means that

$$\left| f\left(\frac{p}{q}\right) \right| = \left| \left(\frac{p}{q} - \alpha\right) \right| |f'(c)| < \frac{1}{q^{n+\varepsilon}}|f'(c)|$$

---

[5]See Appendix F on page 294.

and so

$$\frac{1}{q^n} < \frac{1}{q^{n+\varepsilon}}|f'(c)|.$$

Simplifying results in $q < |f'(c)|^{1/\varepsilon}$.

In both cases the integer q is bounded above and since $|\alpha - p/q| < 1$ it must be that $q(\alpha - 1) < p < q(\alpha + 1)$ and so for each q there are only finitely many values that p can take, resulting in only finitely many fractions p/q once again.

Liouville had the result, but how did he use it to exhibit that first transcendental number?

He constructed the number α as

$$\alpha = \sum_{r=1}^{\infty} \frac{1}{10^{r!}} = \frac{1}{10^{1!}} + \frac{1}{10^{2!}} + \frac{1}{10^{3!}} + \frac{1}{10^{4!}} + \cdots$$

$$= 0.11000100000000000000000001\ldots.$$

Of course it cannot be rational since by its construction its decimal expansion is neither finite nor recurs: it is, then, irrational. Now let us more closely investigate its nature.

Define for $k = 1, 2, 3, \ldots$ the integers p_k, q_k by

$$p_k = 10^{k!}\left(\frac{1}{10^{1!}} + \frac{1}{10^{2!}} + \frac{1}{10^{3!}} + \cdots + \frac{1}{10^{k!}}\right) \quad \text{and} \quad q_k = 10^{k!}.$$

Then

$$\left|\alpha - \frac{p_k}{q_k}\right|$$

$$= \frac{1}{10^{(k+1)!}} + \frac{1}{10^{(k+2)!}} + \frac{1}{10^{(k+3)!}} + \frac{1}{10^{(k+4)!}} \cdots$$

$$= \frac{1}{10^{(k+1)!}}$$

$$\times \left(1 + \frac{1}{10^{k+2}} + \frac{1}{10^{k+2}10^{k+3}} + \frac{1}{10^{k+2}10^{k+3}10^{k+4}} + \cdots\right)$$

$$< \frac{1}{10^{(k+1)!}}\left(1 + \frac{1}{10} + \frac{1}{10^2} + \frac{1}{10^3} + \cdots\right)$$

$$= \frac{\left(\frac{10}{9}\right)}{10^{(k+1)!}} = \frac{\left(\frac{10}{9}\right)}{10^{(k+1)k!}} = \frac{\left(\frac{10}{9}\right)}{(10^{k!})^{k+1}} = \frac{\left(\frac{10}{9}\right)}{(q_k)^{k+1}}.$$

All of which means for each positive integer k

$$\left| \alpha - \frac{p_k}{q_k} \right| < \frac{(\frac{10}{9})}{(q_k)^{k+1}}.$$

Now suppose that α is algebraic of degree n, and consider the inequality in the variable k

$$\frac{(\frac{10}{9})}{(q_k)^{k+1}} < \frac{1}{(q_k)^{n+\varepsilon}},$$

where $\varepsilon > 0$.

This solves to

$$(q_k)^{k+1-n-\varepsilon} > \frac{10}{9}.$$

No matter what value n takes, there is a minimum value of k, and all values bigger than this, such that this inequality holds and therefore such that, for these infinitely many values of k

$$\left| \alpha - \frac{p_k}{q_k} \right| < \frac{1}{(q_k)^{n+\varepsilon}},$$

which is a contradiction to Liouville's theorem. If the number is not algebraic it must be transcendental.

More generally, a Liouville Number α is a number which admits the ultimate accuracy in rational approximation in that the inequality $|\alpha - p/q| < 1/q^k$ is satisfied by infinitely many rationals p/q for all positive integers k. The Liouville Number $\sum_{r=1}^{\infty} 1/10^{r!}$ is just a special case.

The first transcendental number had been discovered. The number was of course entirely contrived to fit with Liouville's Approximation Theorem: it is one thing to manufacture a provably transcendental number (impressive though this feat is) but quite another to decide the nature of the important constants of mathematics such as e and π. Before we consider these great problems, we shall go back a few years to 1840 and once again to the work of Liouville, who took a tentative but elegant stride forward with e.

e, the Non-quadratic Irrational

At the end of the previous chapter we considered an important consequence of numbers which are quadratic irrationals; Liouville proved that, irrational though it is, it is not among them.

He used contradiction, as ever. So, suppose that there are integers a, b, c with $a \neq 0$ for which

$$ae^2 + be + c = 0$$

and rewrite the equation as

$$ae + be + c/e = 0.$$

We can divide e's canonical series expansion into two parts, the first $n + 1$ terms and the remainder, as we have done before, to get

$$e = \sum_{k=0}^{\infty} \frac{1}{k!} = \sum_{k=0}^{n} \frac{1}{k!} + \sum_{k=n+1}^{\infty} \frac{1}{k!}.$$

This guarantees that, when we multiply by $n!$, the number $n!e$ is separated into an integer and a fractional part as

$$n!e = \sum_{k=0}^{n} \frac{n!}{k!} + \sum_{k=n+1}^{\infty} \frac{n!}{k!}.$$

And we can find bounds for that fractional part.
On the one hand,

$$\sum_{k=n+1}^{\infty} \frac{n!}{k!} = \frac{n!}{(n+1)!} + \frac{n!}{(n+2)!} + \frac{n!}{(n+3)!} + \cdots$$

$$> \frac{n!}{(n+1)!} = \frac{1}{n+1}$$

and, on the other,

$$\sum_{k=n+1}^{\infty} \frac{n!}{k!}$$

$$= \frac{1}{n+1} + \frac{1}{(n+2)(n+1)} + \frac{1}{(n+3)(n+2)(n+1)} + \cdots$$

$$< \frac{1}{n+1} + \frac{1}{(n+1)^2} + \frac{1}{(n+1)^3} + \cdots$$

$$= \frac{1/(n+1)}{1 - 1/(n+1)} = \frac{1}{n},$$

which combine to

$$\frac{1}{n+1} < \sum_{k=n+1}^{\infty} \frac{n!}{k!} < \frac{1}{n},$$

which we can rewrite as

$$\sum_{k=n+1}^{\infty} \frac{n!}{k!} = \frac{1}{n+\alpha},$$

where $0 < \alpha < 1$.

All of this means that we can write

$$n!e = I_1 + \frac{1}{n+\alpha},$$

where I_1 is an integer and $0 < \alpha < 1$.

Now we repeat the process for $1/e$ and so $n!/e$, where $1/e = \sum_{k=0}^{\infty}(-1)^k 1/k!$:

$$\frac{n!}{e} = \sum_{k=0}^{n}(-1)^k \frac{n!}{k!} + \sum_{k=n+1}^{\infty}(-1)^k \frac{n!}{k!}.$$

We seek a similar expression for $\sum_{k=n+1}^{\infty}(-1)^k n!/k!$ but matters are a touch more delicate now, with the alternating nature of the series making estimation more difficult, but we can call on the following standard result of real analysis:

Suppose that $S = \sum_{l=0}^{\infty}(-1)^l a_l$ converges, with the $\{a_l\}$ forming a monotonically decreasing sequence tending to 0. If the partial sums are $S_N = \sum_{l=0}^{N}(-1)^l a_l$, then $S_N < S < S_{N+1}$.

To utilize the result in a precise manner we will undertake the change of summation variable $l = k - (n+1)$ in the expression $\sum_{k=n+1}^{\infty}(-1)^k n!/k!$ to transform it to

$$\sum_{k=n+1}^{\infty}(-1)^k \frac{n!}{k!} = \sum_{l=0}^{\infty}(-1)^{l+(n+1)} \frac{n!}{(l+(n+1))!}$$

$$= (-1)^{(n+1)} \sum_{l=0}^{\infty}(-1)^l \frac{n!}{(l+(n+1))!} = (-1)^{(n+1)}S.$$

Using our quoted result above with $N = 1$ we have

$$\frac{1}{n+1} - \frac{1}{(n+2)(n+1)}$$
$$< S < \frac{1}{n+1} - \frac{1}{(n+2)(n+1)} + \frac{1}{(n+3)(n+2)(n+1)}.$$

The left-hand side is evidently $1/(n+2)$ and the right bounded above by $1/(n+1)$. Consequently,

$$\frac{1}{n+2} < S < \frac{1}{n+1}$$

making

$$S = \frac{1}{n+1+\beta},$$

where $0 < \beta < 1$ and so

$$\sum_{k=n+1}^{\infty} (-1)^k \frac{n!}{k!} = (-1)^{(n+1)} S = \frac{(-1)^{(n+1)}}{n+1+\beta},$$

where $0 < \beta < 1$.

All of which means that

$$\frac{n!}{e} = I_2 + \frac{(-1)^{(n+1)}}{n+1+\beta},$$

where I_2 is an integer and $0 < \beta < 1$.

Now multiply our transformed quadratic equation by $n!$ to get

$$a(n!e) + bn! + c(n!/e) = 0$$

and so

$$a\left(I_1 + \frac{1}{n+\alpha}\right) + bn! + c\left(I_2 + \frac{(-1)^{n+1}}{n+1+\beta}\right) = 0,$$

which we rewrite as

$$(aI_1 + bn! + cI_2) + \left(\frac{a}{n+\alpha} + \frac{c(-1)^{n+1}}{n+1+\beta}\right) = 0.$$

Since the first bracket contains an integer, so must the second. Yet we are perfectly free to choose n as we wish and, in particular, we can make it arbitrarily large, which must cause the integer

$$\frac{a}{n+\alpha} + \frac{c(-1)^{n+1}}{n+1+\beta}$$

to become arbitrarily small (taking modulus if necessary); we can only conclude that it must be 0.

We have, then

$$\frac{a}{n+\alpha} + \frac{c(-1)^{n+1}}{n+1+\beta} = 0$$

and so,

$$a(n+1+\beta) + (-1)^{n+1}c(n+\alpha) = 0.$$

Since this is true for all n it must in particular be so for $n = 1, 3$ and this means that

$$a(2+\beta) + c(1+\alpha) = 0 \quad \text{and} \quad a(4+\beta) + c(3+\alpha) = 0.$$

Subtractions yields $a + c = 0$ and substitution back into the first equation $(2+\beta) - (1+\alpha) = 0$ and so $1 + \beta = \alpha$. Since $0 < \beta < 1$, this is a contradiction to the size restriction that $0 < \alpha < 1$.

Liouville easily extended the method to show that e^2 was not a quadratic irrational but further generalization to higher powers of e and generalization to cubic irrationals and above proved elusive. The great question of the transcendence of e required new methods and it was to fall to one of Liouville's students to provide them: Charles Hermite, whose methods set the foundations for those used much later and so effectively by Ivan Niven.

Hermite and e

In fact, in 1873, Hermite proved that e^r is transcendental for all rational $r \neq 0$. We shall look at his remarkable argument for e alone, with the opacity of its inspiration put into suitable perspective by one of his own, and perhaps his most famous, student Henri Poincaré, who said of his teacher:

But to call Hermite a logician! Nothing can appear to me more contrary to the truth. Methods always seemed to be born in his mind in some mysterious way.

Let us look at this particular mysterious method.

Suppose that e is algebraic, in which case it satisfies an equation of lowest degree

$$a_n x^n + a_{n-1} x^{n-1} + a_{n-2} x^{n-2} + \cdots + a_0 = 0$$

with integral coefficients, so $a_0, a_n \neq 0$.

This makes

$$a_n e^n + a_{n-1} e^{n-1} + a_{n-2} e^{n-2} + \cdots + a_0 = 0.$$

For a prime p (whose size we will later choose for our purpose) define a polynomial $f(x)$ of degree $mp + p - 1$ on the interval $0 < x < m$ by

$$f(x) = \frac{x^{p-1}(x-1)^p (x-2)^p \cdots (x-m)^p}{(p-1)!}.$$

With our previously established convention that $f^{(k)}(x) = d^k y/dx^k$, now define the super-function

$$F(x) = f(x) + f^{(1)}(x) + f^{(2)}(x) + \cdots + f^{(mp+p-1)}(x)$$

$$= \sum_{k=0}^{mp+p-1} f^{(k)}(x).$$

With $f(x)$ a polynomial of degree $mp+p-1$ evidently $f^{(k)}(x) = 0$ for $k > mp + p - 1$ and so this stopping point in the definition of $F(x)$ is natural.

We will look to evaluate $F(x)$ at each integer value $r = 0, 1, 2, \ldots, m$ and will deal with two cases separately: $F(r)$ where $r > 0$ and $F(0)$.

First, consider the bracketed terms in the numerator in the definition of $f(x)$. The only way that $f^{(k)}(r) \neq 0$ for $r = 1, 2, \ldots, m$ is when $k = p$, in which case there is a factor of $p!$ in the numerator of the derivative which cancels with the $(p-1)!$ in the denominator to leave p in the numerator. This means that $F(r)$ for $r > 0$ is an integer and has p as a factor.

Now consider $f^{(k)}(0)$. The only way that this is non-zero if $k = p - 1$, in which case, $f^{(p-1)}(0) = (-1)^p \cdots (-m)^p$ and if we choose $p > m$ we can ensure that this term, although an integer, is not divisible by p. This means that $F(0)$ is an integer that does not have p as a factor.

It is clear that

$$F'(x) = f^{(1)}(x) + f^{(2)}(x) + \cdots + f^{(mp+p-1)}(x)$$

since the final term differentiates to 0, which means that

$$F(x) - F'(x) = f(x).$$

Now note that

$$\frac{\mathrm{d}}{\mathrm{d}x}(\mathrm{e}^{-x}F(x)) = \mathrm{e}^{-x}F'(x) - \mathrm{e}^{-x}F(x)$$

$$= -\mathrm{e}^{-x}(F(x) - F'(x)) = -\mathrm{e}^{-x}f(x),$$

which means that

$$\int_0^r \mathrm{e}^{-x}f(x)\,\mathrm{d}x = [-\mathrm{e}^{-x}F(x)]_0^r = F(0) - \mathrm{e}^{-r}F(r).$$

This last equation is commonly known as the *Hermite Identity*. Multiplying by $a_r\mathrm{e}^r$ and summing over r gives

$$\sum_{r=0}^m a_r\mathrm{e}^r \int_0^r \mathrm{e}^{-x}f(x)\,\mathrm{d}x$$

$$= \sum_{r=0}^m a_r\mathrm{e}^r[F(0) - \mathrm{e}^{-r}F(r)] = F(0)\sum_{r=0}^m a_r\mathrm{e}^r - \sum_{r=0}^m a_rF(r)$$

$$= F(0) \times 0 - \sum_{r=0}^m a_rF(r) = -\sum_{r=0}^m a_rF(r)$$

$$= -a_0F(0) - \sum_{r=1}^m a_rF(r).$$

We know that this final right-hand side of the equation is an integer and that the second part of it has the prime p as a common factor. We know also that p does not divide $F(0)$ and that $a_0 \neq 0$, so if we ensure that $p > |a_0|$ we know that p cannot divide a_0 and this makes the first part of the expression a guaranteed fraction when we divide by p, and this means that the whole expression cannot be 0. It is a non-zero integer, then, and its absolute value is $\geqslant 1$.

Now we look at the original left-hand side of the equation. With the bounds on x we have a gross but important upper bound on the size of $f(x)$ with

$$|f(x)| \leqslant \frac{m^{p-1}m^{mp}}{(p-1)!} = \frac{m^{mp+p-1}}{(p-1)!}$$

and so,

$$\left| \sum_{r=0}^{m} a_r e^r \int_0^r e^{-x} f(x)\, dx \right|$$

$$\leqslant \sum_{r=0}^{m} |a_r| e^r \int_0^r e^{-x} |f(x)|\, dx \leqslant \sum_{r=0}^{m} |a_r| e^r \int_0^r e^{-x} \frac{m^{mp+p-1}}{(p-1)!}\, dx$$

$$= \frac{m^{mp+p-1}}{(p-1)!} \sum_{r=0}^{m} |a_r| e^r (1 - e^{-r}) = \frac{m^{mp+p-1}}{(p-1)!} \sum_{r=0}^{m} |a_r| (e^r - 1).$$

With m fixed we can allow p to increase so that the factorial in the denominator reduces the expression to a number less than 1. This is a contradiction to the assumption that e is algebraic.

With this argument fully exposed, Poincaré's view of his teacher is seen to be sustainable, but Hermite was quick to place a disquieting perspective on his achievement when he wrote to the German mathematician Carl Wilhelm Borchardt:

> I shall risk nothing on an attempt to prove the transcendence of π. If others undertake this enterprise, no one will be happier than I in their success. But believe me, it will not fail to cost them some effort.

Lindemann and π

Hermite's comment placed a high bounty on the mathematics necessary to establish π's transcendence and we may suppose that he imagined the need for new techniques to achieve the purpose. We may also suppose that π had been inserted in any number of polynomial equations with integer coefficients by any number of mathematicians in the hope of achieving a contradiction. In the end, and for this purpose *the end* is 1882, it required only a massaging of Hermite's own methods, one that was provided by another German, Ferdinand Lindemann (1852–1939). Whereas Hermite's name remains attached to an impressive collection of mathematical ideas it is for this single result that Lindemann is remembered in the world of research mathematics, one that came about after consultation with Hermite himself and one that called on what is commonly held as the most beautiful formula in the whole of mathematics:

$$e^{i\pi} + 1 = 0.$$

In contrast to Lindemann, the name attached to this is attached to more mathematical and scientific ideas than any other; it is once again that Swiss genius, Leonhard Euler. Yet, it seems not to appear in this explicit form in any of Euler's vast corpus of work and it is also true that others had stated results that imply it, but we find the expression

$$e^{i\theta} = \cos\theta + i\sin\theta$$

in his monumental text *Introductio in Analysin Infinitorum* of 1748, even though he had not evaluated it at $\theta = \pi$.

Since $i = \sqrt{-1}$ is algebraic (as a root of $x^2 + 1 = 0$), it is sufficient to show that $i\pi$ is transcendental since, if π is algebraic, the closure of the algebraic numbers under multiplication would ensure that $i\pi$ is algebraic.

To that end, assume otherwise and let the minimum polynomial having $i\pi$ as a root also have roots $\alpha_1 = i\pi, \alpha_2, \alpha_3, \ldots, \alpha_N$, and suppose that A is its leading coefficient.

Since $e^{i\pi} + 1 = 0$ it must be that

$$(e^{\alpha_1} + 1)(e^{\alpha_2} + 1)(e^{\alpha_3} + 1) \cdots (e^{\alpha_N} + 1) = 0.$$

We need to multiply out these brackets and, to gain a feel for matters, let us take the case $N = 3$, and so generate the $2^3 = 8$

terms

$$(e^{\alpha_1} + 1)(e^{\alpha_1} + 1)(e^{\alpha_1} + 1)$$
$$= e^{\alpha_1} + \alpha_2 + \alpha_3 + e^{\alpha_1} + \alpha_2 + e^{\alpha_2} + \alpha_3$$
$$+ e^{\alpha_1} + \alpha_3 + e^{\alpha_1} + e^{\alpha_2} + e^{\alpha_3} + 1$$
$$= 0.$$

In general, we shall have 2^N powers of e added together, where the powers are of the form $\varepsilon_1 \alpha_1 + \varepsilon_2 \alpha_2 + \varepsilon_3 \alpha_3 + \cdots + \varepsilon_n \alpha_n$, where $\varepsilon_r \in \{0, 1\}$. Write these powers as $\varphi_1, \varphi_2, \varphi_3, \ldots, \varphi_{2^N}$. More than one might be 0 and so we will assume that the first n of them are not so, leaving $q = 2^N - n$ values which are 0. Our multiplied-out expression becomes

$$q + e^{\varphi_1} + e^{\varphi_2} + \cdots + e^{\varphi_n} = 0.$$

We will now define our central polynomial function in much the same way as before.

Recalling that $A\varphi_i$ is an integer for $i = 1, 2, 3, \ldots, n$ the product

$$(x - A\varphi_1)(x - A\varphi_2)(x - A\varphi_3) \cdots (x - A\varphi_n)$$

is a polynomial with integer coefficients; so it must be

$$(Ax - A\varphi_1)(Ax - A\varphi_2)(Ax - A\varphi_3) \cdots (Ax - A\varphi_n)$$
$$= A^n(x - \varphi_1)(x - \varphi_2)(x - \varphi_3) \cdots (x - \varphi_n)$$

and also

$$[A^n(x - \varphi_1)(x - \varphi_2)(x - \varphi_3) \cdots (x - \varphi_n)]^p$$
$$= A^{np}(x - \varphi_1)^p (x - \varphi_2)^p \cdots (x - \varphi_n)^p,$$

where we take p to be an, as yet, unspecified prime number.

Define our polynomial function

$$f(x) = \frac{A^{np} x^{p-1} (x - \varphi_1)^p (x - \varphi_2)^p \cdots (x - \varphi_n)^p}{(p - 1)!}$$

and the super polynomial function

$$F(x) = f(x) + f^{(1)}(x) + f^{(2)}(x) + \cdots + f^{(np+p-1)}(x)$$
$$= \sum_{k=0}^{np+p-1} f^{(k)}(x).$$

Once again,

$$\frac{\mathrm{d}}{\mathrm{d}x}(\mathrm{e}^{-x}F(x)) = -\mathrm{e}^{-x}f(x),$$

which makes

$$\mathrm{e}^{-x}F(x) - F(0) = -\int_0^x \mathrm{e}^{-y}f(y)\,\mathrm{d}y.$$

Making the substitution $y = kx$ in the right hand integral we get

$$\int_0^x \mathrm{e}^{-y}f(y)\,\mathrm{d}y = x\int_0^1 \mathrm{e}^{-kx}f(kx)\,\mathrm{d}k$$

and so

$$\mathrm{e}^{-x}F(x) - F(0) = -x\int_0^1 \mathrm{e}^{-kx}f(kx)\,\mathrm{d}k$$

and finally

$$F(x) - \mathrm{e}^x F(0) = -x\int_0^1 \mathrm{e}^{(1-k)x}f(kx)\,\mathrm{d}k.$$

Now evaluate this expression at each $\varphi_1, \varphi_2, \varphi_3, \ldots, \varphi_n$ and add up the values to get

$$\sum_{j=1}^n [F(\varphi_j) - \mathrm{e}_j^\varphi F(0)] = -\sum_{j=1}^n \varphi_j \int_0^1 \mathrm{e}^{(1-k)\varphi_j}f(k\varphi_j)\,\mathrm{d}k.$$

So,

$$\sum_{j=1}^n F(\varphi_j) - F(0)\sum_{j=1}^n \mathrm{e}^{\varphi_j} = -\sum_{j=1}^n \varphi_j \int_0^1 \mathrm{e}^{(1-k)\varphi_j}f(k\varphi_j)\,\mathrm{d}k.$$

And, if we recall that

$$q + \mathrm{e}^{\varphi_1} + \mathrm{e}^{\varphi_2} + \mathrm{e}^{\varphi_3} + \cdots + \mathrm{e}^{\varphi_n} = q + \sum_{j=1}^n \mathrm{e}^{\varphi_j} = 0$$

we have

$$\sum_{j=1}^n F(\varphi_j) + qF(0) = -\sum_{j=1}^n \varphi_j \int_0^1 \mathrm{e}^{(1-k)\varphi_j}f(k\varphi_j)\,\mathrm{d}k.$$

Let us first look at the left-hand side of the equation.

The non-zero contributions to $F(\varphi_j) = \sum_{k=0}^{np+p-1} f^{(k)}(\varphi_j)$ arise from $k \geqslant p$, where the exponent p will have contributed $p!$ to the numerator, which will cancel with the $(p-1)!$ in the denominator to leave p in the numerator; in short, each $F(\varphi_j)$ is an integer divisible by p.

Now we consider $F(0) = \sum_{k=0}^{np+p-1} f^{(k)}(0)$. Here we have three cases:

$$f^{(k)}(0) = 0 \quad \text{for } k < p - 1,$$
$$f^{(p-1)}(0) = A^{np}(-1)^{np}(\varphi_1\varphi_2\varphi_3 \cdots \varphi_n),$$
$$f^{(k)}(0) = p \times (\text{an integer}) \text{ for the finite number } k > p - 1.$$

In all cases we have an integer and so $F(0)$ is an integer and if we choose p so that it does not divide A or q and sufficiently large so that it does not divide $\varphi_1\varphi_2\varphi_3 \cdots \varphi_n$, we can ensure that p does not divide $qF(0)$: the left-hand side of the above expression is, then, an integer which cannot be 0.

Finally, the right-hand side is again dominated by the $(p-1)!$ on the bottom, with $f(x)$ bounded in the interval $[0, 1]$ and, as before, we can make the expression as small as we please by taking p sufficiently large. Lindemann has brought about the same contradiction as Hermite.

Elementary Observations

Populating the transcendental world beyond these hard-won examples is frustratingly difficult.

Certainly, if $x \neq 0$ is algebraic and y is transcendental, then $x + y$ and xy are transcendental: these ensure the transcendence of the likes of $\pi + \sqrt{2}$ and $\sqrt{2}e$. But unfortunately, unlike the algebraics and like the irrationals, the transcendentals are not closed under addition and multiplication so, although we have π and e as transcendentals, the nature of the likes of $\pi \pm e$, πe, π/e, π^e, π^π, e^e remain unknown.

In fact, we have no idea whether $\pi + e$ and πe are even irrational but we can use a simple observation to show that at least one of them is so: the quadratic equation whose roots are π and e is $x^2 - (\pi + e)x + \pi e = 0$ and if both $\pi + e$ and πe were rational then π and e would be algebraic: we now know that they are not!

We can make other elementary observations using the closure of the algebraics to gain a little more ground: not both of $\pi \pm e$ can be algebraic since this would make their sum 2π algebraic and not both of πe and π/e can be algebraic since that would make their product, π^2 algebraic and, with π transcendental, so must be its powers. And with this thought in mind recall that, with the Zeta function from chapter 5, $\zeta(2n) = C\pi^{2n}$, where C is rational: the nature of it for odd integers is a nightmare but is clearly transcendental for even integers.

The Transcendental Cascade

The above observations and others like them enable some inroads to be made into the question of the transcendence of numbers but there is one result and its sequel which stand apart in their significance. The reader may have noticed an obvious omission from the above list of combinations of e and π: e^π. It is missing because it is known to be transcendental. It is known to be transcendental because of a celebrated result known as the *Gelfond Schneider Theorem* and that came about in answer to a question in the most famous list of mathematical questions ever compiled.

We have mentioned German mathematical leviathan David Hilbert (1862–1943) before: in 1900 the 38 year old gave what is generally regarded as one of the most outstanding expository lectures of all time. Given at the Sorbonne in Paris at the Second International Congress of Mathematicians, the structure of *The Problems of Mathematics* broke from established tradition by concentrating on future mathematical challenge rather than summarizing past endeavor; at its heart is a list of 23 problems, some specific, some general, all of which he considered paramount to the future advance of mathematics, 10 of which he spoke about in his lecture. Its seductive and inspirational opening paragraph is:

> Who of us would not be glad to lift the veil behind which the future lies hidden; to cast a glance at the next advances of our science and at the secrets of its development during future centuries? What particular goals will there be toward which the leading mathematical spirits of coming generations will strive? What new methods and new facts in the wide and rich field of mathematical thought will the new centuries disclose?

In his preliminary comments he drew attention to the fruitfulness of the theretofore unsuccessful attempts at proving Fermat's Last Theorem and in 1928, in a talk given to a lay audience, he selected this great question and two of his numbered problems for comparison in difficulty: Problem 8, which is now universally known as the Riemann Hypothesis, and Problem 7, where he asked about the nature of the number $2^{\sqrt{2}}$ and more generally of α^β, where $\alpha \neq 0, 1$ is algebraic and β an algebraic irrational. Specifically, he asked whether such a number always represents a transcendental or at least an irrational number.

In that 1928 talk Hilbert expressed his belief that the Riemann Hypothesis would be resolved in his lifetime, Fermat's Last Theorem within the lifetime of the younger members of the audience but that *no one in this room will live to see a proof of the Seventh*. The great man was wrong. At the time of printing, the Riemann Hypothesis remains unproved; Fermat's Last Theorem succumbed to Andrew Wiles in 1994, when a doddery survivor of that talk could have read about that particular drama unfolding, and the Seventh Problem was to be fully resolved merely six years after the talk, in 1934. Hermann Weyl was later to refer to anyone who solved one of the 23 problems as someone who had entered *the honours class of mathematicians* and the names of the Russian Aleksandr Gelfond and the German Theodor Schneider were destined to be members of that most distinguished group as they, independently and simultaneously, were to provide the first complete resolution of Problem 7.

In fact, the first significant assault on Hilbert's Problem 7 was an extension of an earlier result of Gelfond, when in 1929, the year following Hilbert's popular talk, he established the following result:

> If $\alpha \neq 0, 1$ is algebraic and $\beta = i\sqrt{b}$, where $b > 0$ is rational, with $i^2 = -1$, then α^β is transcendental.

With this we have that $2^{i\sqrt{2}}$ is transcendental. Removing i from the result fell informally to the German Carl Siegel (whose work was seminal in this and numerous other areas of number theory), but who decided not to publish his result; his magnanimous view was that Gelfond would soon see how to extend the result himself. In fact, the name of the Russian mathematician R. O. Kuzman is attached to the first appearance in print of the result in 1930, but

Gelfond was to return in 1934 (along with Schneider) with the full resolution of Hilbert's seventh problem with

> If $\alpha \ne 0, 1$ and β are algebraic, with $\beta \notin \mathbb{Q}$, then α^{β} is transcendental.

Now, if we use the Euler identity $e^{i\pi} + 1 = 0$ we get

$$e^{\pi} = (e^{i\pi})^{-i} = (-1)^{-i}$$

and we have what we need. At no extra cost we have the stronger result that the square root is transcendental too, since

$$\sqrt{e^{\pi}} = e^{\pi/2} = (e^{i\pi/2})^{-i} = i^{-i}.$$

Lindemann had stated but not proved a stronger result than the one he proved above, indicating that he thought it could be proved using similar ideas to those he had used. Its statement:

> Given any distinct algebraic numbers $\alpha_1, \alpha_2, \alpha_3, \ldots, \alpha_n$, if $a_1 e^{\alpha_1} + a_2 e^{\alpha_2} + a_3 e^{\alpha_3} + \cdots + a_n e^{\alpha_n} = 0$ for algebraic numbers $a_1, a_2, a_3, \ldots, a_n$ then $a_1 = a_2 = a_3 = \cdots = a_n = 0$.

It was to be his contemporary and fellow countryman, Karl Weierstrass (1815–97), who was to supply the (far from easy) proof to bring about what has become known as the *Lindemann–Weierstrass Theorem* and which has wide consequences:

- e^{α} is transcendental for any non-zero algebraic number α: if e^{α} were algebraic we could write $e^{\alpha} = -a_1/a_2$ for some non-zero algebraic numbers a_1, a_2 and so $a_1 + a_2 e^{\alpha} = a_1 e^0 + a_2 e^{\alpha} = 0$ and we have $a_1 e^{\alpha_1} + a_2 e^{\alpha_2} = 0$, where $\{\alpha_1, \alpha_2\} = \{0, \alpha\}$ are two distinct algebraic numbers.
- π is transcendental: since $e^{i\pi} + 1 = 0$, if $i\pi$ were algebraic we would have $a_1 e^{\alpha_1} + a_2 e^{\alpha_2} = 1 \times e^{i\pi} + 1 \times e^0 = 0$ for the two algebraic numbers $\{a_1, a_2\} = \{1, 1\}$ and $\{\alpha_1, \alpha_2\} = \{\pi, 0\}$. Since $i\pi$ is transcendental, so must π be.
- $\sin \alpha$, $\cos \alpha$, $\tan \alpha$, $\sinh \alpha$, $\cosh \alpha$, $\tanh \alpha$ are transcendental for any non-zero algebraic number α, as are their inverse functions: these succumb through their definitions, identities and/or applications of the result. For example, we could argue that, since $\sin^2 \alpha + \cos^2 \alpha = 1$, if either of the two functions is algebraic the other must be and so must be

$\cos \alpha + i \sin \alpha = e^{i\alpha}$ and this is transcendental. Alternatively, this last identity gives $\sin \alpha = (e^{i\alpha} - e^{-i\alpha})/(2i)$ and so $e^{i\alpha} - e^{-i\alpha} - 2i \sin \alpha = 0$ and if $\sin \alpha$ were algebraic we would have $1 \times e^{\alpha_1} + (-1) \times e^{\alpha_2} + (-2i \sin \alpha) \times e^{\alpha_3} = 0$, where $\{a_1, a_2, a_3\} = \{1, -1, -2i \sin \alpha\}$ and $\{\alpha_1, \alpha_2, \alpha_2\} = \{i\alpha, -i\alpha, 0\}$.

In what is otherwise something of a mire of mathematical ignorance, there is something of a surprising strength of knowledge arising from this single result; more can be gained from it but with these statements we hope to have conveyed sufficient of its influence.

The Problem of Constructability

With the transcendence of π assured, not only had a highly resistant mathematical citadel been breached but the last of three geometric problems of antiquity had been resolved:

1. duplication of the cube (the Delian Problem);
2. trisection of an arbitrary angle;
3. squaring the circle.

These succinct descriptions of problems that have challenged for millennia expand to the possibility, using straight edge and compass alone, of constructing:

1. the length of the side of a cube whose volume is twice that of a given (unit) cube, and so a length $\sqrt[3]{2}$;
2. the line trisecting any given angle;
3. the side of a square whose area is that of a given (unit) circle, and so a length $\sqrt{\pi}$.

We will not enter into the detailed history of these problems but their antiquity and fame may be judged by the inclusion of the third in the comedy *Birds*, written by the Greek playwright Aristophanes and performed in 414 B.C.E.[6]; the plot involved the founding of a city in the sky by the main character, Peisthetaerus, with the welcome help of birds and the unwelcome intervention of some humans. One such human was the geometer Meton

---

[6]At least in some translations of the work. Others would have it that there was to be a circle in a square, but we have chosen a recognized authority of stature for the translation.

(who actually existed as such) and whose character, replete with straight edge and compass, offered a set of plans for the city's design accompanied by the words:[7]

> ...will measure it with a straight measuring rod, having applied it, that your circle may become four-square; and in the middle of it there may be a market place, and that there may be straight roads leading to it, to the very centre...

to which Peisthetaerus replies:

> The fellow's a Thales!

To have squared the circle would have been a feat worthy of that great sage whom we mentioned in chapter 1.

In fact, the resolution of the three problems took around 2,200 years and took place in two parts, separated by under 50 years. With these pages strewn with a litany of famous mathematical names, Pierre Wantzel (1814–48) rather stands out, but this Frenchman owes his place here as the first person to publish proofs that the duplication of the cube and the trisection of an arbitrary angle are both impossible; also, his absence would deprive us of a character, if not a name, of note since, according to his mathematical collaborator Jean Claude Saint-Venant:

> He was blameworthy for having been too rebellious to the counsels of prudence and of friendship. Ordinarily he worked evenings, not lying down until late; then he read, and took only a few hours of troubled sleep, making alternately wrong use of coffee and opium, and taking his meals at irregular hours until he was married. He put unlimited trust in his constitution, very strong by nature, which he taunted at pleasure by all sorts of abuse.

Death at the age of 34 hardly seems surprising in spite of the implied balancing effect of his wife but it is his work as a 23 year old that interests us here, for it was in 1837 that he published in that prestigious *Journal de Mathématiques Pures et Appliquées*, to which we have already alluded in this chapter, his analysis of the

---

[7]Translation by W. J. Hickie, available from Google Books at http://books.google.com/books?id=Cm4NAAAAYAAJ&source=gbs_navlinks_s.

two problems – and more besides.[8] Later, in 1845, he was also to give a new proof of the impossibility of solving the general quintic by radicals.

His methods, now subsumed into Galois Theory, are rather general in that they attack the wider problem of straight edge and compass constructability but our specific needs are catered for by an observation drawn from them that the cubic equation with rational coefficients

$$x^3 + ax^2 + bx + c = 0$$

has a root which is a constructible number[9] only if it has a root which is a rational number:[10] consequently, no rational roots implies no constructible roots.

The duplication of the cube requires a length of $\sqrt[3]{2}$ to be constructible and this is a root of $x^3 - 2 = 0$, which we know has no rational roots.

The impossibility of the trisection of an angle of 60° requires the standard identity

$$\cos 3\theta = 4\cos^3\theta - 3\cos\theta$$

with $3\theta = 60°$.

Putting $x = \cos 20°$ into the equation results in

$$\tfrac{1}{2} = 4x^3 - 3x, \quad x^3 - \tfrac{3}{4}x - \tfrac{1}{8} = 0 \quad \text{or} \quad 8x^3 - 6x - 1 = 0.$$

Again, there are no rational roots and so no constructible roots: $\cos 20°$ is therefore not constructible and so the angle of 60° cannot be trisected.

In passing, also in that 1837 paper, Wantzel filled a gap left by Gauss, who had famously shown that a regular p-gon can be constructed (with straight edge and compass) if p is a prime of the form $2^{2^n} + 1$. Gauss had warned readers not to attempt the construction for primes other than these but gave no proof

---

[8]Pierre Wantzel, 1837, Recherche sur les moyens de reconnaître si un problème de géométrie peut se résoudre à la règle et au compas, *Journal de Mathématiques Pures et Appliquées* 2:366–72.

[9]That is, capable of being constructed with straight edge and compass.

[10]For a proof the reader may wish to consult, see, for example, Craig Smorynski, 2007, *History of Mathematics: A Supplement* (Springer).

that this was impossible; Wantzel demonstrated that the regular heptagon is incapable of construction which, in terms of the
approach above, translates to a degree 7 equation which can easily
be reduced to the cubic equation

$$x^3 + x^2 - 2x - 1 = 0$$

with its lack of rational roots.

Wantzel's methods did not extend to the problem of squaring
the circle and nor did any others until Lindemann's proof that π is
transcendental; since all constructible numbers are algebraic, the
impossibility is assured. A great and ancient problem had been
solved by Lindemann and one wonders whether he found it galling
to be told:

> What good is your beautiful investigation regarding π? Why
> study such problems, since irrational numbers do not exist.

Such was the view of his influential and significant countryman
Leopold Kronecker (1823–91), whose critical views were to have a
destructive effect on the next contributor to the development of
irrational numbers.

Cantor and Infinity

Kronecker described the German Georg Cantor (1845–1918) as *a
corrupter of youth* and Poincaré referred to his set theory, with
its transfinite numbers and all of its counterintuitive nature and
paradox, *a malady that would one day be cured*. It would not
have contributed to concord when in 1874, just a year after Hermite's proof appeared, Cantor published a paper which showed in
his precise manner that, although transcendental numbers might
be fiercely elusive, there are more of them than there are algebraic numbers. To achieve this remarkable result he needed to
utilize his notion of the relative sizes of infinite sets, with a *countable set* the smallest such and defined as one which can be put
into one-to-one correspondence with $\mathbb{N}^+ = \{1, 2, 3, \dots\}$: another
way of describing a countable set is that it is one which can be
written down as a list, with the place in the list of each element
determined by that one-to-one correspondence.

If the algebraic numbers can be shown to be countable and the
real numbers $\mathbb{R}$ not so, the complement of the algebraics in the

Table 7.1.

Height h	Value of n	Polynomial	Real roots
2	1	$x = 0$	0
3	1	$2x = 0, x \pm 1 = 0$	$0, \pm 1$
	2	$x^2 = 0$	
4	1	$3x = 0, 2x \pm 1 = 0, x \pm 2 = 0$	$0, \pm\frac{1}{2},$
	2	$2x^2 = 0, x^2 \pm 1 = 0, x^2 \pm x = 0$	$\pm 1, \pm 2$
	3	$x^3 = 0$	
5	1	$4x = 0, 3x \pm 1 = 0,$	$0, \pm\frac{1}{3}, \pm\frac{1}{2},$
		$2x \pm 2 = 0, x \pm 3 = 0$	$\pm 1, \pm 3, \pm\frac{1}{\sqrt{2}},$
	2	$3x^2 = 0, 2x^2 \pm 1 = 0, x^2 \pm 2 = 0$	$\pm\sqrt{2}, \pm 2,$
		$2x^2 \pm x = 0, x^2 \pm 2x = 0,$	$\frac{\pm 1 \pm \sqrt{5}}{2}$
		$x^2 \pm x \pm 1 = 0$	
	3	$2x^3 = 0, x^3 \pm 1 = 0,$	
		$x^3 \pm x = 0, x^3 \pm x^2 = 0$	
	4	$x^4 = 0$	

reals must be more numerous than the algebraics themselves: there are more transcendentals than algebraics.

His arguments appeared in *J. De Crelle* in 1874 in a paper which marked the beginning of the subject that we now know as set theory and he combined two of them to achieve the purpose. We will reproduce them in largely their original, if expanded, form.

First, the algebraic numbers are countable.

Every real algebraic number is a root of a polynomial equation with integer coefficients:

$$a_n x^n + a_{n-1} x^{n-1} + a_{n-2} x^{n-2} + \cdots + a_0 = 0.$$

Define the *height* of the polynomial to be

$$h = n + |a_1| + |a_2| + |a_3| + \cdots + |a_n|,$$

then $h \geqslant 2$ is an integer. Every polynomial has a height and table 7.1 lists the first few polynomials by height and also the real roots generated by them. Evidently, there are only a finite number of polynomials of any given height, each of which has only finitely many roots. This means that there are only finitely many real algebraic numbers arising from polynomials of any given height. Now we can list the algebraic numbers by the height of their defining polynomials, omitting repetition and ordering by size for each height, as below:

$$(0), (-1, 1), (-2, -\tfrac{1}{2}, \tfrac{1}{2}, 2),$$
$$(-3, \tfrac{-1-\sqrt{5}}{2}, -\sqrt{2}, -\tfrac{1}{\sqrt{2}}, \tfrac{1-\sqrt{5}}{2}, -\tfrac{1}{3}, \tfrac{1}{3}, \tfrac{-1+\sqrt{5}}{2}, \tfrac{1}{\sqrt{2}}, \sqrt{2}, \tfrac{1+\sqrt{5}}{2}, 3), \dots.$$

This listing of the algebraic numbers means that they must be countable. Now Cantor needed $\mathbb{R}$, the set of real numbers, not to be so. His subtle approach to this allowed him at once to provide a contradiction to the assumption that $\mathbb{R}$ is countable but also a precise method of generating irrationals and transcendentals. It relies on several intrinsic properties of any interval $[\alpha, \beta]$ of the real numbers:

- it is linearly ordered:
- it is dense: that is, between any two numbers in the interval there is another number;
- it has no gaps: that is, if it is partitioned into two nonempty sets A and B in such a way that, using the ordering, every member of A is less than every member of B, then there is a boundary point, x, so that every element less than or equal to x is in A and every element greater than x is in B.

Now we shall consider his statement and his proof of it.

> Given any sequence of real numbers S and any interval $[\alpha, \beta]$ on the real line, it is possible to determine a number η in $[\alpha, \beta]$ that does not belong to S. Hence, one can determine infinitely many such numbers η in $[\alpha, \beta]$.

For, suppose that the sequence of real numbers is

$$S = \{\omega_1, \omega_2, \omega_3, \dots\}.$$

Look along S to find the first two numbers in it which lie within $[\alpha, \beta]$; call the smaller number α_1 and the larger β_1. Now consider these two as endpoints of the nested interval $[\alpha_1, \beta_1]$ and continue along S to find the first two numbers which lie within $[\alpha_1, \beta_1]$; call the smaller number α_2 and the larger β_2 and consider these as endpoints of the nested interval $[\alpha_2, \beta_2]$. Continue the process to generate a succession of nested intervals,

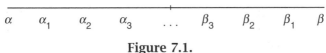

Figure 7.1.

the endpoints of which are numbers from the sequence S. There are two possibilities: the nesting finishes after a finite number of iterations or it continues indefinitely.

In the first case it must be that eventually it is only possible to find at most one number, ω, from S which lies in the final interval $[\alpha_N, \beta_N]$: take η to be any number in $[\alpha_N, \beta_N]$ other than α_N, β_N or ω and we are guaranteed that this η does not lie in S.

In the second case there are an infinite number of nested intervals with end points each forming an infinite sequence $\{\alpha_n\}$ and $\{\beta_n\}$ of numbers from S, the first increasing and bounded above (by β) and the second decreasing and bounded below (by α). Each sequence must, then, have a limit point in S, which Cantor wrote as α_∞ and β_∞ respectively. Once again, there are two cases:

$\alpha_\infty = \beta_\infty$: take η to be this common limit, which by its construction cannot be in S;

$\alpha_\infty < \beta_\infty$: take η to be any number in the interval $[\alpha_\infty, \beta_\infty]$, which, again by construction, cannot be in S.

In all cases we have a number in the interval $[\alpha, \beta]$ which cannot lie in the countable sequence S.

As a diagram we have figure 7.1.

Under the assumption that the real numbers are countable, take $S = \mathbb{R}$ and we have the contradiction that shows otherwise. Put these two last results together and we have the algebraic numbers a countable subset of the uncountable real numbers, which means that the transcendental numbers are uncountable and so 'far more numerous' than the algebraics.

Cantor's task had been accomplished with this non-constructive argument. Yet, the argument's essence lies with the general nature of the sequence S; it is any (infinite) sequence of real numbers and his argument shows that all such sequences lack some real number. It becomes a constructive one with his additional observation in that paper that for any sequence of algebraic numbers the case $\alpha_\infty = \beta_\infty$ pertains: take any such S and we are guaranteed to home in on a real number that is not contained within it.

For example, if we take S to be the (countable) set of algebraic numbers his argument does not yield a contradiction, but a real number that is not a member of the sequence; that is, a number that is not algebraic – that is, a transcendental number. To suit our purpose we need an ordering of the algebraics but Cantor has provided this in his first argument and if we extract in order those (say) in the interval $[0, 1]$ (which means that the sequence begins with $\{0, 1, \frac{1}{2}, \frac{1}{3}, \frac{-1+\sqrt{5}}{2}, \frac{1}{\sqrt{2}}, \ldots\}$) the procedure can be used to generate a transcendental number between 0 and 1. Unfortunately, the programming difficulties are great but Robert Gray[11] has tackled them using a computationally more tractable method of ordering the algebraics, not by the *height* of the defining polynomial but by what he calls the *size* of the polynomial, defined using standard notation as $\max\{n, a_0, a_1, a_2, \ldots, a_n\}$, with a secondary ordering to arrange those having the same size. The magnitude of the computational challenge is apparent from his findings that, using the previous notation,

$$\alpha_7 = \omega_{1,406,370} = 0.57341146\ldots$$

and

$$\beta_7 = \omega_{1,057,887} = 0.57341183\ldots.$$

The transcendental number being generated begins with

$$0.573411\ldots$$

and perhaps it is of interest to know that the defining polynomial equations for these two algebraic approximations are, respectively,

$$x^6 - x^5 + 2x^4 + 3x^3 - x^2 + x - 1 = 0$$

and

$$x^6 - 4x^5 - x^4 + 5x^3 + 2x^2 + 3x - 3 = 0.$$

Modest but, we hope, illuminating is the process applied to the rational numbers, where we will choose the convenient ordering by denominator

$$\frac{1}{2}, \frac{1}{3}, \frac{2}{3}, \frac{1}{4}, \frac{2}{4}, \frac{3}{4}, \frac{1}{5}, \frac{2}{5}, \frac{3}{5}, \frac{4}{5}, \frac{1}{6}, \frac{2}{6}, \frac{3}{6}, \frac{4}{6}, \frac{5}{6}, \frac{1}{7}, \ldots$$

All rational numbers in the interval $(0, 1)$, with some repetition, appear.

---

[11] Robert Gray, 1994, Georg Cantor and transcendental numbers, *American Mathematical Monthly* 101:819–32.

Table 7.2.

		Fractional		Decimal	
n	n	α_n	β_n	α_n	β_n
2	1	$\dfrac{1}{3}$	$\dfrac{1}{2}$	0.33333333...	0.5
8	18	$\dfrac{2}{5}$	$\dfrac{3}{7}$	0.4	0.42857142...
127	60	$\dfrac{7}{17}$	$\dfrac{5}{12}$	0.411176470...	0.41666666...
390	797	$\dfrac{12}{49}$	$\dfrac{17}{41}$	0.41379310...	0.41463414...
4794	2375	$\dfrac{41}{99}$	$\dfrac{29}{70}$	0.41414141...	0.41428571...
14,098	28,302	$\dfrac{70}{169}$	$\dfrac{99}{239}$	0.41420118...	0.41422594...
165,839	82,790	$\dfrac{239}{577}$	$\dfrac{169}{408}$	0.41421143...	0.41421568...
484,044	968,713	$\dfrac{408}{985}$	$\dfrac{577}{1393}$	0.41421319...	0.41421392...
5,651,234	2,824,861	$\dfrac{1393}{3363}$	$\dfrac{985}{2378}$	0.41421349...	0.41421362...

Using Cantor's arguments, we are bound to home in on a real number between 0 and 1 that is not in the sequence; that is an irrational number. The programming task is trivial and the computational task quite manageable, yielding table 7.2. The reader will note the linkage between numerators and denominators in the α_n and β_n, which may be a temptation for investigation – and also that $\sqrt{2} - 1 = 0.41421356\ldots$: also, sequence A084068 of the Online Encyclopedia of Integer Sequences[12] may be of interest.

A Hierarchy of (Ir)rationality

We will leave this chapter on transcendental numbers, far from its end, but with a quantitative division of the real numbers into the three types: rational, algebraic and transcendental, and also a quantitative distinction between the relative transcendence of transcendental numbers.

---

[12]http://www.research.att.com/~ njas/sequences/index.html.

Table 7.3.

α	μ_α
Rational	1
Algebraic	2
Liouville	∞
e	2
π	8.016045...
$\zeta(2)$	5.441243...

The distinction is brought about by considering the Liouville bounds on approximation in the following manner:

If we return to our generic inequality $|\alpha - p/q| < 1/q^{\mu+\varepsilon}$ we have seen that, if α is rational, there are only finitely many rational approximants to $|\alpha - p/q| < 1/q^{1+\varepsilon}$ for any $\varepsilon > 0$; if α is algebraic, Roth's result shows that there are only finitely many rational approximants to $|\alpha - p/q| < 1/q^{2+\varepsilon}$ for any $\varepsilon > 0$. In both cases the type of number has been squeezed as far as possible in its capability of approximation. With transcendental numbers we only know that they are capable of approximation for $\mu \geqslant 2$ and it is therefore reasonable to ask for any particular number how much μ can be increased beyond 2, still with that infinite number of rational approximations.

With such thoughts we define the irrationality measure for a given number α to be the upper bound μ_α of μ for which the inequality has an infinite number of solutions for any $\varepsilon > 0$.

The most extreme case is that of the Liouville numbers, for which $\mu_\alpha = \infty$. As we move to other numbers matters become deeply technical and very difficult. Very little is known about specific numbers and there are still fewer general theorems. Table 7.3 lists a few of the latest upper bounds at the time of going to press: notice that $\mu_e = 2$ is no contradiction (algebraic $\Rightarrow \mu = 2$, but $\mu = 2 \not\Rightarrow$ algebraic) and that the appearance of $\zeta(3)$ is consistent with the number almost certainly being transcendental.

A general theorem that has been established is a typically frustrating one: this last result is one of many of the Russian Aleksandr Khinchin, who gave a non-existential proof that almost all real numbers have an irrationality measure of 2. We have echoes of Cantor.

Continued Fractions Revisited

If you change the way you look at things, the things you look at change.

Max Planck

In chapter 3 we considered the application of continued fractions in the eighteenth century, as they were used to grind out the first proofs of the irrationality of e and π. Their role in the study of irrational numbers is far greater though, with a number of the results of previous chapters implicitly dependent on them; here we make some of that dependence explicit.

We shall continue to restrict our interest to simple continued fractions and so to finite or infinite expressions of the form

$$\alpha = a_0 + \cfrac{1}{a_1 + \cfrac{1}{a_2 + \cfrac{1}{a_3 + \cdots}}}$$

and we will adopt the notation that the nth convergent of the continued fraction is written

$$\frac{p_n}{q_n} = [a_0; a_1, a_2, a_3, \ldots, a_n].$$

We will first dispose of the finite case, and in doing so rational numbers, with the following two results.

1. If two simple continued fractions are equal,

$$[a_0; a_1, a_2, a_3, \ldots, a_n] = [b_0; b_1, b_2, b_3, \ldots, b_m],$$

and $a_n, b_m > 1$ (a technicality needed to cope with a non-serious ambiguity in the final digit of the representation), then $m = n$ and $a_1 = b_1, a_2 = b_2, a_3 = b_3, \ldots, a_n = b_m$.

2. A number is rational if and only if its simple continued fraction expansion is finite.

These we will not prove, but their veracity is easily established by simple proofs that are part of elementary theory and easily found.

With these we move to irrational numbers and so to the infinite continued fractions that can be used to represent them.

First, a result of Euler in the eighteenth century and Lagrange in the nineteenth combine to an important categorization, which again we choose not to prove:

> The continued fraction representation of a number is periodic if and only if the number is a quadratic surd.

We discussed quadratic surds in chapter 3. Periodic continued fractions, just as with periodic decimals, are those that at some stage repeat themselves. For example,

$$[0; 1, 2, 3, 4, 5, 6, 4, 5, 6, 4, 5, 6, \ldots] = \frac{2557 - \sqrt{18229}}{1690}.$$

We shall return to quadratic irrationals soon, but first we will look at the general case, wherein we have the necessarily infinite continued fraction representation of an irrational number α, without regard to any pattern in its digits.

The Hierarchy of the Convergents

First, the continued fraction process can be conveniently represented[1] using matrix multiplication in that, with our established notation,

$$\begin{pmatrix} p_n & p_{n-1} \\ q_n & q_{n-1} \end{pmatrix} = \begin{pmatrix} a_0 & 1 \\ 1 & 0 \end{pmatrix} \begin{pmatrix} a_1 & 1 \\ 1 & 0 \end{pmatrix} \begin{pmatrix} a_2 & 1 \\ 1 & 0 \end{pmatrix} \cdots \begin{pmatrix} a_n & 1 \\ 1 & 0 \end{pmatrix},$$

which means that

$$\begin{pmatrix} p_n & p_{n-1} \\ q_n & q_{n-1} \end{pmatrix} = \begin{pmatrix} p_{n-1} & p_{n-2} \\ q_{n-1} & q_{n-2} \end{pmatrix} \begin{pmatrix} a_n & 1 \\ 1 & 0 \end{pmatrix}$$

---

[1] The result is susceptible to proof by induction.

and this means that

$$p_n = a_n p_{n-1} + p_{n-2} \quad \text{and} \quad q_n = a_n q_{n-1} + q_{n-2},$$

from which we can see that both numerator and denominator of the convergents are increasing with n.

These relationships can be developed into the interrelationships between the p_ns and q_ns in that

$$p_{n+1} q_n - p_n q_{n+1} = (-1)^n,$$
$$p_{n+1} q_{n-1} - p_{n-1} q_{n+1} = (-1)^{n+1} a_{n+1}.$$

Our interest is with the first of the two, which we will all but prove.

The left side, as a function of n, can write in terms of $n - 1$ in that

$$\begin{aligned} p_{n+1} q_n - p_n q_{n+1} &= (a_{n+1} p_n + p_{n-1}) q_n - p_n (a_{n+1} q_n + q_{n-1}) \\ &= a_{n+1} p_n q_n + p_{n-1} q_n - a_{n+1} p_n q_n - p_n q_{n-1} \\ &= p_{n-1} q_n - p_n q_{n-1} \\ &= -(p_n q_{n-1} - p_{n-1} q_n). \end{aligned}$$

Observe now that $p_0 = a_0$, $q_0 = 1$, $p_1 = a_0 a_1 + 1$, $q_1 = a_1$ and we can quickly check the initial truth of the statement and we have all we need for an inductive proof; alternatively, the use of the multiplicative nature of determinants can be utilized to give the result.

This result discloses three things:

1. The greatest common divisors,

$$[p_n, q_n] = [p_{n+1}, p_n] = [q_{n+1}, q_n] = 1.$$

2. The form

$$\frac{p_{r+1}}{q_{r+1}} - \frac{p_r}{q_r} = \frac{(-1)^r}{q_r q_{r+1}}$$

shows that the convergents approximate α alternatively from above and below.

3. Adding this form gives

$$\sum_{r=0}^{n} \left(\frac{p_{r+1}}{q_{r+1}} - \frac{p_r}{q_r} \right) = \sum_{r=0}^{n} \frac{(-1)^r}{q_r q_{r+1}}$$

and the series on the left cascades to

$$\frac{p_{n+1}}{q_{n+1}} - \frac{p_0}{q_0} = \sum_{r=0}^{n} \frac{(-1)^r}{q_r q_{r+1}}$$

and so we have

$$\frac{p_{n+1}}{q_{n+1}} = \frac{p_0}{q_0} + \sum_{r=0}^{n} \frac{(-1)^r}{q_r} q_{r+1}.$$

From this last, and with the q_ns known to be increasing, we conclude

$$\frac{p_2}{q_2} = \frac{p_0}{q_0} + \left(\frac{1}{q_0 q_1} - \frac{1}{q_1 q_2}\right) = \frac{p_0}{q_0} + \frac{1}{q_1}\left(\frac{1}{q_0} - \frac{1}{q_2}\right) > \frac{p_0}{q_0},$$

$$\frac{p_4}{q_4} = \frac{p_2}{q_2} + \left(\frac{1}{q_2 q_3} - \frac{1}{q_3 q_4}\right) = \frac{p_2}{q_2} + \frac{1}{q_3}\left(\frac{1}{q_2} - \frac{1}{q_4}\right) > \frac{p_2}{q_2},$$

and it is clear that we can continue this to

$$\frac{p_0}{q_0} < \frac{p_2}{q_2} < \frac{p_4}{q_4} < \cdots .$$

Similarly with the odd convergents

$$\frac{p_3}{q_3} = \left(\frac{p_0}{q_0} + \frac{1}{q_0 q_1}\right) + \frac{1}{q_2 q_3} - \frac{1}{q_1 q_2} = \frac{p_1}{q_1} + \frac{1}{q_2 q_3} - \frac{1}{q_1 q_2}$$

$$= \frac{p_1}{q_1} + \frac{1}{q_2}\left(\frac{1}{q_3} - \frac{1}{q_1}\right) < \frac{p_1}{q_1},$$

which continues to

$$\frac{p_1}{q_1} > \frac{p_3}{q_3} > \frac{p_5}{q_5} > \cdots .$$

Finally, we link the odd and the even convergents with

$$\frac{p_{2n+1}}{q_{2n+1}} = a_0 + \sum_{r=0}^{2n} \frac{(-1)^r}{q_r q_{r+1}} \quad \text{and} \quad \frac{p_{2n}}{q_{2n}} = a_0 + \sum_{r=0}^{2n-1} \frac{(-1)^r}{q_r q_{r+1}}$$

to get

$$\frac{p_{2n+1}}{q_{2n+1}} - \frac{p_{2n}}{q_{2n}} = \left(a_0 + \sum_{r=0}^{2n} \frac{(-1)^r}{q_r q_{r+1}}\right) - \left(a_0 + \sum_{r=0}^{2n-1} \frac{(-1)^r}{q_r q_{r+1}}\right)$$

$$= \frac{(-1)^{2n}}{q_r q_{r+1}} = \frac{1}{q_r q_{r+1}} > 0$$

Table 8.1.

n	p_n/q_n	n	p_n/q_n
0	3.0000000000000000000	1	3.1428571428571428571
2	3.1415094339622641509	3	3.1415929203539823008
4	3.1415926530119026040	5	3.1415926539214210447
6	3.1415926534674367055	7	3.1415926536189366233
8	3.1415926535810777712	9	3.1415926535514039784
10	3.1415926535893891715	11	3.1415926535898153832
12	3.1415926535897926593	13	3.1415926535897934025
14	3.1415926535897931602	15	3.1415926535897932578
16	3.1415926535897932353	17	3.1415926535897932390
18	3.1415926535897932383	19	3.1415926535897932384

and we have the full hierarchy

$$\frac{p_0}{q_0} < \frac{p_2}{q_2} < \frac{p_4}{q_4} < \cdots < \alpha < \cdots < \frac{p_5}{q_5} < \frac{p_3}{q_3} < \frac{p_1}{q_1}.$$

So the even-numbered convergents form a monotone increasing sequence converging to α and the odd-numbered convergents form a monotone decreasing sequence converging to α. Table 8.1 demonstrates the behaviour for $\alpha = \pi$.

Note particularly that the path left–right–left... provides an oscillating sequence and that the biggest even convergent is smaller than the smallest odd convergent – and that π lies between them, as it does between every consecutive pair of them.

Rational Approximation Revisited

In chapter 6 we repeatedly pressed for ever greater accuracy of the rational approximation to a given irrational number, and we judged an approximation p/q to be better than p'/q' provided that $|q\alpha - p| < |q'\alpha - p'|$. As we squeezed the tolerance so we sieved the irrationals, finishing with the Lagrange Spectrum of 'break-points', which is intimately connected to the exotic Markov Spectrum of numbers. Our proofs were non-existential and disguised the fact that continued fractions usually lay at their heart, and we will here take the opportunity to revisit some of them, couched in this alternative language.

The convergents p_n/q_n are the best possible approximations to a number in the sense that if p and q are integers with $q < q_{n+1}$ then $|q\alpha - p| \geqslant |q_n\alpha - p_n|$.

We will prove this and do so by assuming that, on the contrary, $|q\alpha - p| < |q_n\alpha - p_n|$, where $q < q_{n+1}$.

If we define the two numbers u and v by the equations

$$p = up_n + vp_{n+1} \quad \text{and} \quad q = uq_n + vq_{n+1}.$$

Then u and v are integers, since,

$$\begin{pmatrix} p \\ q \end{pmatrix} = \begin{pmatrix} p_n & p_{n+1} \\ q_n & q_{n+1} \end{pmatrix} \begin{pmatrix} u \\ v \end{pmatrix}$$

and so

$$\begin{pmatrix} u \\ v \end{pmatrix} = \frac{1}{p_nq_{n+1} - p_{n+1}q_n} \begin{pmatrix} q_{n+1} & -p_{n+1} \\ -q_n & p_n \end{pmatrix} \begin{pmatrix} p \\ q \end{pmatrix}$$

$$= (-1)^{n+1} \begin{pmatrix} q_{n+1} & -p_{n+1} \\ -q_n & p_n \end{pmatrix} \begin{pmatrix} p \\ q \end{pmatrix},$$

which means that

$$u = (-1)^{n+1}(pq_{n+1} - p_{n+1}q) \quad \text{and} \quad v = (-1)^{n+1}(-pq_n + p_nq).$$

Furthermore, $v \neq 0$, since if it were, $p = up_n$, $q = uq_n$ and we would have the contradiction

$$|q\alpha - p| = |uq_n\alpha - up_n| = |u||q_n\alpha - p_n| \geqslant |q_n\alpha - p_n|.$$

Also, $u \neq 0$, since if it were, $q = |v|q_{n+1}$, which contradicts $q < q_{n+1}$.

Further, u and v must be of opposite signs since, if $v < 0\ up_n = p - vp_{n+1} \Rightarrow u > 0$, whereas, if $v > 0$, $q < vq_{n+1} \Rightarrow uq_n < 0 \Rightarrow u < 0$.

Also note that

$$\alpha q_n - p_n = q_n\left(\alpha - \frac{p_n}{q_n}\right) \quad \text{and} \quad \alpha q_{n+1} - p_{n+1} = q_{n+1}\left(\alpha - \frac{p_{n+1}}{q_{n+1}}\right)$$

must be of opposite signs because, from observation (3) on page 213, α lies between p_n/q_n and p_{n+1}/q_{n+1}. Combine these results to the observation that these two must have the same sign:

$$u(\alpha q_n - p_n) \quad \text{and} \quad v(\alpha q_{n+1} - p_{n+1}).$$

But then

$$|q\alpha - p| = |\alpha(uq_n + vq_{n+1}) - (up_n + vp_{n+1})|$$
$$= |u(\alpha q_n - p_n) + v(\alpha q_{n+1} - p_{n+1})| \geqslant |\alpha q_n - p_n|,$$

which is, once again, a contradiction.

We also argued that any irrational number can be approximated so that $|\alpha - p/q| < 1/(2q^2)$. These next two results show precisely how.

If $|\alpha - p/q| < 1/(2q^2)$ then p/q is a convergent of the continued fraction of α.

To see this, let p_n/q_n be some convergent of α and, with the best approximation result above, consider

$$\left|\frac{p}{q} - \frac{p_n}{q_n}\right| = \left|\left(\frac{p}{q} - \alpha\right) + \left(\alpha - \frac{p_n}{q_n}\right)\right| \leqslant \left|\frac{p}{q} - \alpha\right| + \left|\alpha - \frac{p_n}{q_n}\right|$$
$$= \frac{1}{q}|\alpha q - p| + \frac{1}{q_n}|\alpha q_n - p_n|$$
$$\leqslant \frac{1}{q}|\alpha q - p| + \frac{1}{q_n}|\alpha q - p|$$
$$= \left(\frac{1}{q} + \frac{1}{q_n}\right)|\alpha q - p| < \left(\frac{1}{q} + \frac{1}{q_n}\right)\frac{1}{2q}.$$

Write the spectrum of denominators of the convergents of α as $\{q_1, q_2, q_3, \ldots\}$ and choose n so that $q_n \leqslant q < q_{n+1}$, in which case, $1/q \leqslant 1/q_n$ and we have

$$\left|\frac{p}{q} - \frac{p_n}{q_n}\right| < \frac{2}{q_n} \times \frac{1}{2q} = \frac{1}{qq_n},$$

which means that the positive integer $|pq_n - qp_n| < 1$ and so it must be that $p/q = p_n/q_n$.

For any irrational number α, and for any two consecutive convergents, $p_n/q_n, p_{n+1}/q_{n+1}$ to α, either $|\alpha - p_n/q_n| < 1/(2q_n^2)$ or $|\alpha - p_{n+1}/q_{n+1}| < 1/(2q_{n+1}^2)$ or both.

Now we have

$$\left|\frac{p_n}{q_n} - \frac{p_{n+1}}{q_{n+1}}\right| = \frac{|p_n q_{n+1} - p_{n+1} q_n|}{q_n q_{n+1}} = \frac{1}{q_n q_{n+1}}.$$

Using the result that α lies between two consecutive convergents, we have

$$\left|\alpha - \frac{p_n}{q_n}\right| + \left|\frac{p_{n+1}}{q_{n+1}} - \alpha\right| = \left|\frac{p_{n+1}}{q_{n+1}} - \frac{p_n}{q_n}\right| = \frac{1}{q_n q_{n+1}},$$

so

$$\left|\alpha - \frac{p_n}{q_n}\right| + \left|\alpha - \frac{p_{n+1}}{q_{n+1}}\right| = \frac{1}{q_n q_{n+1}}.$$

But

$$(a - b)^2 = a^2 + b^2 - 2ab > 0$$

so

$$ab < \tfrac{1}{2}(a^2 + b^2).$$

Take $a = 1/q_n$ and $b = 1/q_{n+1}$, then

$$\frac{1}{q_n}\frac{1}{q_{n+1}} = \frac{1}{q_n q_{n+1}} < \frac{1}{2}\left(\frac{1}{q_n^2} + \frac{1}{q_{n+1}^2}\right) = \frac{1}{2q_n^2} + \frac{1}{2q_{n+1}^2},$$

so

$$\left|\alpha - \frac{p_n}{q_n}\right| + \left|\alpha - \frac{p_{n+1}}{q_{n+1}}\right| < \frac{1}{2q_n^2} + \frac{1}{2q_{n+1}^2}$$

this means that either or both of these inequalities hold

$$\left|\alpha - \frac{p_n}{q_n}\right| < \frac{1}{2q_n^2} \quad \text{or} \quad \left|\alpha - \frac{p_{n+1}}{q_{n+1}}\right| < \frac{1}{2q_{n+1}^2}.$$

As we pressed further we met the result of Hurwitz, which in part showed that the 2 can be increased to $\sqrt{5}$. Again, the (and his) use of partial fractions reveals why, with the result:

For any 3 consecutive convergents of the irrational number α, p_n/q_n, p_{n+1}/q_{n+1}, p_{n+2}/q_{n+2} at least one must satisfy $|\alpha - p/q| < 1/(\sqrt{5}q^2)$.

For, suppose not, then

$$\left|\alpha - \frac{p_n}{q_n}\right| \geqslant \frac{1}{\sqrt{5}q_n^2} \quad \text{and} \quad \left|\alpha - \frac{p_{n+1}}{q_{n+1}}\right| \geqslant \frac{1}{\sqrt{5}q_{n+1}^2},$$

so

$$\frac{1}{q_n q_{n+1}} = \left| \alpha - \frac{p_n}{q_n} \right| + \left| \alpha - \frac{p_{n+1}}{q_{n+1}} \right| \geqslant \frac{1}{\sqrt{5} q_n^2} + \frac{1}{\sqrt{5} q_{n+1}^2},$$

so

$$\frac{1}{\sqrt{5}} \left(\frac{q_{n+1}}{q_n} \right) + \frac{1}{\sqrt{5}} \left(\frac{q_n}{q_{n+1}} \right) \leqslant 1.$$

Write $\lambda = q_{n+1}/q_n$ and we have $\lambda + 1/\lambda \leqslant \sqrt{5}$. Since λ is rational we have $\lambda + 1/\lambda < \sqrt{5}$. So,

$$\lambda^2 - \sqrt{5}\lambda + 1 = (\lambda - \tfrac{\sqrt{5}+1}{2})(\lambda - \tfrac{\sqrt{5}-1}{2}) < 0.$$

This means that $\frac{\sqrt{5}-1}{2} < \lambda < \frac{\sqrt{5}+1}{2}$ and in particular $\lambda < \frac{\sqrt{5}+1}{2}$, which makes $1/\lambda > \frac{\sqrt{5}-1}{2}$.

Write $\mu = q_{n+2}/q_{n+1}$ and $\mu < \frac{\sqrt{5}+1}{2}$. But $q_{n+2} = a_{n+2} q_{n+1} + q_n$ so $q_{n+2}/q_{n+1} = a_{n+2} + q_n/q_{n+1} \geqslant 1 + q_n/q_{n+1}$:

$$\mu \geqslant 1 + \frac{1}{\lambda} > 1 + \frac{\sqrt{5}-1}{2} = \frac{\sqrt{5}+1}{2}.$$

A contradiction.

Now let us see how the use of continued fractions and this result of Hurwitz can be combined to give a measure of irrationality.

The World's Most Irrational Number

We know from this version of Hurwitz's theorem that for every irrational number there are an infinite number of rational approximants satisfying $|\alpha - p/q| < 1/(\sqrt{5}q^2)$ and that we should look to the continued fraction expansion of α for the best among them. Among this infinite number of ever more accurate approximants we can seek any for which the Hurwitz bound of the error is large compared with the actual error of the approximation. Consider, for example, table 8.2, which lists the first four convergents to π which are within the Hurwitz-prescribed error margin: and note the second row of data. We see that the error in the approximation $\frac{355}{133}$ is far smaller than the guaranteed error margin and we can quantify this discrepancy using the ratio of the two, which appears in the final column of the table. With this we can make precise the idea one number being 'more irrational' than another:

Table 8.2.

n	p_n/q_n	Error (E)	$(1/\sqrt{5}q_n^2)(H)$	E/H
1	$\dfrac{22}{7}$	1.26499×10^{-3}	9.12681×10^{-3}	0.13854
3	$\dfrac{355}{113}$	2.66764×10^{-7}	3.50234×10^{-5}	0.0076167
5	$\dfrac{104348}{33215}$	3.31628×10^{-10}	4.05365×10^{-10}	0.818097
7	$\dfrac{312689}{99532}$	2.91434×10^{-11}	4.51429×10^{-11}	0.64558

Table 8.3.

n	p_n/q_n	Error (E)	$(1/\sqrt{5}q_n^2)(H)$	E/H
1	3	0.281718	0.447214	0.629941
4	$\dfrac{19}{7}$	3.99611×10^{-3}	9.12681×10^{-3}	0.437844
7	$\dfrac{193}{71}$	2.80307×10^{-5}	8.87153×10^{-5}	0.315963
10	$\dfrac{2721}{1001}$	1.10177×10^{-7}	4.46321×10^{-7}	0.246857

α is deemed more irrational than β if the minimum of the E/H ratio for α is greater than that for β. We judge that approximations with E/H near 0 are tighter than we might expect, whereas those near 1 are barely creeping into compliance with the Hurwitz result. Table 8.3 gives the data for e and table 8.4 that for the Golden Ratio.

It is looking as though e might be harder to approximate than π – and that the Golden Ratio might be more resistant still, with the relative error barely keeping below the upper limit of 1 and seemingly approaching it. In the front matter we hinted that the Golden Ratio is, in fact, the *most* irrational number: with the stage now set, we will prove it so. To be precise we will show that, with the continued fraction approximations,

$$\left| \varphi - \frac{p_n}{q_n} \right| \xrightarrow{n \to \infty} 1 \times \frac{1}{\sqrt{5}q_n^2}$$

Table 8.4.

n	p_n/q_n	Error (E)	$(1/\sqrt{5}q_n^2)(H)$	E/H
1	3	0.381966	0.447214	0.854102
3	$\dfrac{5}{3}$	0.0486327	0.0496904	0.978714
5	$\dfrac{13}{8}$	6.96601×10^{-3}	6.98771×10^{-3}	0.996894
7	$\dfrac{34}{21}$	1.01363×10^{-3}	1.01409×10^{-3}	0.999547

or, put another way,

$$q_n^2 \left| \varphi - \frac{p_n}{q_n} \right| = q_n |q_n\varphi - p_n| \xrightarrow{n \to \infty} \frac{1}{\sqrt{5}}.$$

To begin the proof, we know that

$$p_n = F_{n+1} \quad \text{and} \quad q_n = F_n.$$

And also we know the Binet (or Euler or De Moivre) formula for the nth Fibonnaci number in terms of φ:

$$F_n = \frac{\varphi^n - (1 - \varphi)^n}{\sqrt{5}}.$$

Now consider

$$q_n^2 \left| \varphi - \frac{p_n}{q_n} \right|$$

$$= q_n |q_n\varphi - p_n|$$

$$= \frac{\varphi^n - (1 - \varphi)^n}{\sqrt{5}} \left| \left(\frac{\varphi^n - (1 - \varphi)^n}{\sqrt{5}} \right) \varphi - \frac{\varphi^{n+1} - (1 - \varphi)^{n+1}}{\sqrt{5}} \right|$$

$$= \frac{\varphi^n - (1 - \varphi)^n}{\sqrt{5}} \left| \frac{\varphi^{n+1} - \varphi(1 - \varphi)^n}{\sqrt{5}} - \frac{\varphi^{n+1} - (1 - \varphi)^{n+1}}{\sqrt{5}} \right|$$

$$= \frac{\varphi^n - (1 - \varphi)^n}{\sqrt{5}} \left| \frac{(1 - \varphi)^{n+1} - \varphi(1 - \varphi)^n}{\sqrt{5}} \right|$$

$$= \frac{\varphi^n - (1 - \varphi)^n}{\sqrt{5}} |(1 - \varphi)^n| \left| \frac{(1 - \varphi) - \varphi}{\sqrt{5}} \right|$$

$$= \frac{\varphi^n - (1 - \varphi)^n}{\sqrt{5}} |(1 - \varphi)^n| \left| \frac{1 - 2\varphi}{\sqrt{5}} \right|.$$

Since $\varphi = \frac{1+\sqrt{5}}{2}$, $1 - 2\varphi = -\sqrt{5}$, and so

$$q_n^2\left|\varphi - \frac{p_n}{q_n}\right| = \frac{\varphi^n - (1-\varphi)^n}{\sqrt{5}}|(1-\varphi)^n|\left|\frac{-\sqrt{5}}{\sqrt{5}}\right|$$

$$= \frac{\varphi^n - (1-\varphi)^n}{\sqrt{5}}|(1-\varphi)^n|$$

$$= \frac{|[\varphi(1-\varphi)]^n| - (1-\varphi)^n|(1-\varphi)^n|}{\sqrt{5}}$$

recalling that the defining equation for φ is $\varphi^2 = \varphi + 1$ we have $\varphi(1-\varphi) = -1$ and so

$$q_n^2\left|\varphi - \frac{p_n}{q_n}\right| = \frac{|(-1)^n| - (1-\varphi)^n|(1-\varphi)^n|}{\sqrt{5}}$$

$$= \frac{1 - (1-\varphi)^n|(1-\varphi)^n|}{\sqrt{5}} \xrightarrow{n\to\infty} \frac{1-0}{\sqrt{5}} = \frac{1}{\sqrt{5}}$$

and we are done.

That Markov Spectrum

The second part of Hurwitz's result showed that an essential boundary had been reached with, for example, φ being incapable of being approximated to a finer tolerance; from this came the cascade of boundary irrational numbers and from them came the mysterious spectrum of numbers (and those equivalent to them) investigated by Markov. This list began

$$\frac{-1+\sqrt{5}}{2}, \quad -1+\sqrt{2}, \quad \frac{-9+\sqrt{221}}{14}, \quad \frac{-23+\sqrt{1517}}{38}, \quad \ldots$$

We know from an earlier result that the continued fraction form of these quadratic irrationals will be recurring; let us observe these first four:

$$\frac{-1+\sqrt{5}}{2} = \{0; 1, 1, 1, 1, 1, 1, 1, 1, 1, 1, 1, 1, 1,$$
$$1, 1, 1, 1, 1, 1, 1, 1, 1, 1, 1, 1, 1, 1, 1, 1, \ldots\},$$

$$-1+\sqrt{2} = \{0; 2, 2, 2, 2, 2, 2, 2, 2, 2, 2, 2, 2, 2,$$
$$2, 2, 2, 2, 2, 2, 2, 2, 2, 2, \ldots\},$$

Table 8.5.

α	Boundary numbers
$[0;\bar{1}]$	$\sqrt{5}$
$[0;\bar{2}]$	$\sqrt{8}$
$[0;\overline{2_2,1_2}]$	$\dfrac{\sqrt{221}}{5}$
$[0;\overline{2_2,1_4}]$	$\dfrac{\sqrt{1517}}{13}$
$[0;\overline{2_4,1_2}]$	$\dfrac{\sqrt{7565}}{29}$
$[0;\overline{2_2,1_6}]$	$\dfrac{\sqrt{2600}}{17}$
$[0;\overline{2_2,1_8}]$	$\dfrac{\sqrt{71285}}{89}$
$[0;\overline{2_6,1_2}]$	$\dfrac{\sqrt{257045}}{169}$
$[0;\overline{2_2,1_2,2_2,1_4}]$	$\dfrac{\sqrt{84680}}{97}$

$$\frac{-9+\sqrt{221}}{14} = \{0;2,2,1,1,2,2,1,1,2,2,1,1,2,$$
$$2,1,1,2,2,1,1,2,2,1,1,\dots\},$$

$$\frac{-23+\sqrt{1517}}{38} = \{0;2,2,1,1,1,1,2,2,1,1,1,1,2,$$
$$2,1,1,1,1,2,2,1,1,1,1,\dots\},$$

and a pattern is discerned which is absent in their surd form: periodic they certainly are, but it is this special nature of the periodicity that attracts the eye. Table 8.5 lists these and a few more in their continued fraction form, together with the associated boundary numbers of each: for brevity we adopt the standard convention that, for example, $[0;1,1,1,1,2,2,1,1,1,1,2,2,1,1,1,$ $12,2,\dots] = [0;\overline{1_4,2_2}]$.

With this we are closer to the mystery of the Markov numbers but revealing it fully remains beyond our reach. We can, though,

make one further incursion into the theory. The boundary numbers were not isolated but grouped into equivalence classes with
the equivalence relation:[2]

> α and α' are equivalent if there are integers such that $\alpha' = (a\alpha + b)/(c\alpha + d)$, where $|ad - bc| = 1$.

For example, the following are equivalent:

$$\sqrt{8} = [2; 1, 4, 1, 4, 1, 4, 1, 4, 1, 4, 1, 4, 1, 4, 1, 4, 1, 4, 1,$$
$$4, 1, 4, 1, 4, 1, 4, 1, 4, 1, 4, \ldots],$$

$$\frac{7\sqrt{8} + 5}{4\sqrt{8} + 3} = [1; 1, 2, 1, 2, 1, 4, 1, 4, 1, 4, 1, 4, 1, 4, 1, 4, 1, 4, 1, 4, \ldots],$$

$$\frac{11\sqrt{8} + 7}{8\sqrt{8} + 5} = [1; 2, 1, 1, 1, 2, 1, 4, 1, 4, 1, 4,$$
$$1, 4, 1, 4, 1, 4, 1, 4, 1, 4, 1, 4, 1, 4, 1, 4, 1, 4, 1, \ldots].$$

It takes but a second to notice that the last two expansions look
to be identical with the first after the first few terms. In fact, the
continued fraction statement is

> $\alpha \sim \alpha'$ if and only if their continued fraction expansions are
> identical from some point on

and, in addition,

> a/c and b/d are successive convergents in the continued
> expansion expression for α' provided that $\alpha > 1$ and $c > d > 0$.

For example, the list of convergents in the last case begins

$$1, \quad \frac{3}{2}, \quad \frac{4}{3}, \quad \frac{7}{5}, \quad \frac{11}{8}, \quad \frac{29}{21}, \quad \frac{40}{29}, \quad \ldots$$

There is more – much more – to expose in the study of irrational
numbers through the use of continued fractions, but we hope that
the reader will have gained some feeling for this most fruitful
interaction. In the next chapter we will move to quite a different
aspect of irrationality – for which continued fractions show a clear
limitation.

---

[2]See Appendix E on page 289.

CHAPTER NINE

The Question and Problem of Randomness

Anyone who considers arithmetical methods of producing random digits is, of course, in a state of sin.

John von Neumann

We will move from the question, how *irrational* is an irrational number, to another: how *random* is the decimal expansion of an irrational number? Our measure of irrationality allows comparisons to be made, but what measure is there of randomness? In the above quotation, the genius von Neumann provides a hint of the complexities involved when we attempt to harness randomness and we shall expose some of the difficulties in what follows. First, we shall once again eliminate the rationals.

Characterizing Rationals by Decimal Expansions

This is achieved by the combination of three results:

- If the decimal expansion of a number is finite it is rational: for example,

$$3.14159 = 3\frac{14159}{10000}$$

and, in general,

$$a_0.a_1a_2a_3\ldots a_n = a_0\frac{a_1a_2a_3\cdots a_n}{10^n}.$$

- If the decimal expansion of a number is recurring it is also rational: for example,

$$3.\overline{14159} = 3 + 0.14159 + 0.0000014159$$
$$+ 0.000000000014159 + \cdots$$
$$= 3 + 14159(10^{-5} + 10^{-10} + 10^{-15} + \cdots)$$
$$= 3 + 14159 \times \frac{10^{-5}}{1 - 10^{-5}}$$
$$= 3 + \frac{14159}{10^5 - 1} = 3\frac{14159}{99999}.$$

In general,

$$a_0.\overline{a_1 a_2 a_3 \ldots a_n}$$
$$= a_0 + a_1 a_2 a_3 \cdots a_n(10^{-n} + 10^{-2n} + 10^{-3n} + \cdots)$$
$$= a_0 + a_1 a_2 a_3 \cdots a_n \times \frac{10^{-n}}{1 - 10^{-n}}$$
$$= a_0 \frac{a_1 a_2 a_3 \cdots a_n}{10^n - 1}.$$

- Conversely, if a number is rational, its decimal expansion is either finite of recurring. To find the decimal expansion of a/b, where $a < b$ and $b > 1$, with $r_0 = a$:

$$r_0 = b \times a_1 + r_1, \quad 0 \leqslant r_1 \leqslant b - 1,$$
$$r_1 = b \times a_2 + r_2, \quad 0 \leqslant r_2 \leqslant b - 1,$$
$$r_2 = b \times a_3 + r_3, \quad 0 \leqslant r_3 \leqslant b - 1,$$
$$\vdots$$

The $a_1, a_2, a_3, \ldots$ are the digits of the decimal expansion of a/b. If any of the remainders is 0, the expansion is finite and anyway, since $0 \leqslant \{r_0, r_1, r_2, \ldots\} \leqslant b - 1$, using the Pigeonhole Principle mentioned on page 162, it must be that among the integers $r_0, r_1, r_2, \ldots, r_b$ that some $r_i = r_j$. In which case,

$$a_{i+1} = a_{j+1}, \quad r_{i+1} = r_{j+1}, \quad a_{i+2} = a_{j+2}, \quad r_{i+2} = r_{j+2}, \quad \ldots$$

and so the decimal expansion is periodic.

Struggling with Randomness

So, with irrational numbers we have guaranteed infinite, non-recurring decimal expansions: but do they form a sequence of *random numbers*? Whatever careful meaning we might give to the term *random decimal expansion*, we should not be happy to include 0.01 001 0001 00001 000001 ... as an example of one. The number of 0s increase by 1 at each stage and so the decimal expansion is neither finite nor recurring; the number must be irrational. Yet, it has nothing of *randomness* about it with its clear prescription for generation and its use of just two digits: we would hardly be content to use a section of it to form a random selection of the digits from $1, 2, 3, \ldots, 9$.

The number is a pathology, constructed for purpose to make a point. To avoid such a thing surely we need, at least, an even mix of the digits $1, 2, 3, \ldots, 9$; but then

0.12345678911223344556677889

991112223334445556667778889999...

is one such and a second irksome pathology. Again it is assuredly an example of an irrational number but as surely lacks the intuitive quality of *randomness*.

So, perhaps we need all of the ten digits represented on an equal basis and that they be mixed together in a ... *random* manner.

Let us take a natural irrational number: π. Its decimal expansion starts with the 650 digits:

3.1415926535 8979323846 2643383279 5028841971 6939937510
5820974944 5923078164 0628620899 8628034825 3421170679
8214808651 3282306647 0938446095 5058223172 5359408128
4811174502 8410270193 8521105559 6446229489 5493038196
4428810975 6659334461 2847564823 3786783165 2712019091
4564856692 3460348610 4543266482 1339360726 0249141273
7245870066 0631558817 4881520920 9628292540 9171536436
7892590360 0113305305 4882046652 1384146951 9415116094
3305727036 5759591953 0921861173 8193261179 3105118548
0744623799 6274956735 1885752724 8912279381 8301194912
9833673362 4406566430 8602139494 6395224737 1907021798
6094370277 0539217176 2931767523 8467481846 7669405132
0005681271 4526356082 7785771342 7577896091 7363717872

continuing forever[1] in its 'patternless way'. So, we imagine, does the decimal expansion of e. Might the two be patternless in precisely the same manner, that is, might the decimal expansion of π end with that of e? Impossible, surely. Assume it true, then

$$\pi = 3.14159\ldots a_n 27181828\ldots$$
$$= 3.14159\ldots a_n + 10^{-(n+1)} \times 2.7181828\ldots$$

and so

$$\pi = \frac{p}{q} + 10^{-(n+1)} \times \text{e} \quad \text{and} \quad \pi - 10^{-(n+1)} \times \text{e} = \frac{p}{q}.$$

All of which means that $10^{(n+1)}\pi - \text{e} = p/q$. We have mentioned before that it is currently unknown whether either of $\pi \pm \text{e}$ is rational; in fact, our degree of ignorance is still greater: there is no pair of integers m, n for which it is known whether or not $m\pi + n\text{e}$ is irrational. We may not want to bet on it but there is no present contradiction to the assumption and π might end in e!

Whatever the case, as the digits of π are generated, there is hardly a feeling of what we intuitively take to be randomness about the production of the next one in the sequence; each digit is precisely prescribed, there being no doubt, for example, that the millionth such[2] is 1. The eye senses a nice balance between the digits, though: the decimal expansion *looks* random. If it is so, surely all sequences of digits of all lengths will appear somewhere in the expansion. Of course, to begin a test of this requires a (presumably very long) truncated approximation for us to search and a small exercise in binomial probability reveals that, if the decimal expansion is assumed random, then the probability of a sequence of length n digits appearing in its $N > n$ digit truncation is

$$1 - (1 - 0.1^n)^{N-n+1}.$$

With this we require $N \approx 2.4 \times 10^{10}$ to have a 90% chance of a sequence of ten digits (say, 0123456789 or the 10-digit ISBN of this book) appearing: in fact, 0123456789 appears for the first

---

[1]At the time of writing the record for its decimal expansion is held by Fabrice Bellard with nearly 2700 billion digits.

[2]After the decimal point.

time starting at the 17,387,594,880th digit; whereas 0691143420 continues to prove elusive.[3]

We could play with such ideas indefinitely but the fact is that it is just that: play. Randomness is a very, very subtle concept with its study properly belonging to statisticians more than mathematicians. In 1951 Dick Lehmer, a pioneer in computing and especially computational number theory, disclosed an instructive view:

> A random sequence is a vague notion in which each term is unpredictable to the uninitiated and whose digits pass a certain number of tests traditional with statisticians and depending somewhat on the uses to which the sequence is to be put.

Two examples in particular set the modern standard for such tests, each assuming the numbers have been written in binary form. The first is the National Institute of Standards and Technology (NIST) Test Suite, a statistical package of 16 tests:

1. the frequency (monobit) test;
2. frequency test within a block;
3. the runs test;
4. test for the longest-run-of-1s in a block;
5. the binary matrix rank test;
6. the discrete fourier transform (spectral) test;
7. the non-overlapping template matching test;
8. the overlapping template matching test;
9. maurer's 'universal statistical' test;
10. the Lempel–Ziv compression test;
11. the linear complexity test;
12. the serial test;
13. the approximate entropy test;
14. the cumulative sums (cusums) test;
15. the random excursions test;
16. the random excursions variant test.

And a second accepted approach, the Diehard test suite of George Marsaglia:

---

[3]With this level of truncation the 13-digit ISBN would appear with a likelihood of about 0.2%.

1. birthday spacings;
2. overlapping permutations;
3. ranks of matrices;
4. monkey tests;
5. count the 1s;
6. parking lot test;
7. minimum distance test;
8. random spheres test;
9. the squeeze test;
10. overlapping sums test;
11. runs test;
12. the craps test.

There is much that is mysterious here and we will leave the interested reader to unravel matters as they wish; the simple fact is that the definition of a random sequence simply does not properly exist and all that does exist are sets of reasonable criteria the passing of which cannot confer the name random to a number but make it plausible that it is so – whatever that means.

Normality

In 1909 the French mathematician Emil Borel introduced the concept of *normal* numbers as a precise mathematical way of characterizing our intuitive idea of randomness in a number's expansion. Our interest has been, and will remain, with decimal expansions but the concept is base dependent and naturally divides into three parts[4]:

- A number x is *simply normal* in base b if each digit in its base b expansion appears with a frequency of $1/b$.
- A number x is *normal in base b* if each digit and sequence of digits in its base b expansion appear with a frequency determined by the length of the sequence. That is, single digits with a frequency $1/b$, pairs of digits with a frequency of $1/b^2$, triplets of digits with a frequency of $1/b^3$, etc.
- A number is *normal* if it is normal in every base $b \geqslant 2$.

---

[4]There are alternative definitions.

We see that the idea accords with intuition, not as a definition of random but as a necessary condition for such a quality to be attached.

Simply normal numbers are those whose digits appear with equal frequency, not that this precludes arbitrarily long sequences of arbitrary distributions of frequencies: table 9.1 shows the unsurprising digit distribution of the first 50,000,000,000 digits of π. Yet simple normality is insufficient for irrationality; take, for example, the rational number

$$0.01234567890123456789012345678 9\ldots.$$

It is also possible to construct numbers which are simply normal in one base but not another. For example, the following simply normal binary number converts to a rational decimal number:

$$0.10101010\ldots = 2^{-1} + 2^{-3} + 2^{-5} + \cdots$$

$$= \frac{2^{-1}}{1 - 2^{-2}} = \frac{2}{3} = 0.6666\ldots.$$

The concept of normal in base b brings with it more, in that every sequence of digits must appear in the expansion of the number an infinite number of times. Such numbers cannot, therefore, be rational. Notice also that, if a number is normal in base b, it must also be normal in bases b^k for all $k \geqslant 1$ (and the reverse is true too). Normal numbers in a definite sense embody the concept of randomness.

So, with a precise definition to work with, can we find normal numbers? The answer is *yes* but not the ones we want: the first example was given by Waclaw Sierpiński in 1916 and another important one in 1975 by Gregory Chaitin; they are exotic, to say the least, and constructed for purpose. It is widely conjectured that the 'natural' transcendental numbers, including π, are normal but proving the conjecture is currently out of reach. We do not even know if any of the obvious constants are simply normal in any base; specifically, we do not know if there are an infinite numbers of 2s in the decimal expansion of π. This said, recall that Cantor had demonstrated the uncountability of the elusive transcendentals; parallel with this, in that 1909 paper, Borel had proved that almost all real numbers are normal!

There is, though, a celebrated number which is normal in base 10. It was constructed by the Englishman David Champernowne (1912–2000) and appeared in a paper[5] written as a Cambridge undergraduate in 1933. It requires an understanding of the Cambridge University Tripos system to appreciate fully Champernowne's achievement in taking a double First in the Mathematics Tripos one year early and then, having changed subjects, obtaining another First in Part II of the Economics Tripos: perhaps it is enough to realize that it was Keynes who influenced him to change from mathematics to economics and that he was not only a close friend of but also intellectually comparable with Alan Turing.

The number, Champernowne's Constant, is the decimal expansion consisting of all non-negative integers concatenated in sequence

0.1 2 3 4 5 6 7 8 9 10 11 12 13 14 15 16 17 18 19 20 21 ....

Moreover, the number is known to be transcendental.

We saw in the last chapter how useful continued fraction representation can be in studying irrational numbers, but Champernowne's Constant provides us with a telling example of an irrational number which is resistant to analysis using them: its continued fraction form is[6]

[0; 8, 9, 1, 149083, 1, 1, 1, 4, 1, 1, 1, 3, 4, 1, 1, 1, 15, 457540
1113910310 7648364662 8242956118 5996039397
1045755500 0662004393 0902626592 5631493795
3207747128 6563138641 2093755035 5209460718
3089984575 8014698631 4883359214 1783010987,
6, 1, 1, 21, 1, 9, 1, 1, 2, 3, 1, 7, 2, 1, 83, 1, 156, 4, 58, 8, 54, ...].

The reader should look carefully at the placing of the separating commas between the digits in the expression; the number after 15 has 166 digits and the one following 54 will have 2504 digits, which means that this particular irrational number is *very* well

[5]D. G. Champernowne, 1933, The construction of decimals normal in the scale of ten, *J. Lond. Math. Soc.* 8:254–60.

[6]http://www.research.att.com/~njas/sequences/A030167.

Table 9.1.

Digit	Frequency
0	5000012647
1	4999986263
2	5000020237
3	4999914405
4	5000023598
5	4999991499
6	4999928368

approximated by rationals.[7] Taking the pleasantly large 149083, for example, yields the very accurate rational approximation

$$0.1234567891011\cdots - \frac{1490839}{12075796} \sim 4.4 \times 10^{-15}.$$

And then there is the number constructed by A. S. Besicovitch. In 1934 he proved that the irrational number, now known as Besicovitch's Constant, formed by concatenating the squares of the integers is normal in base 10:

$$0.149162536496481100121144169196\ldots.$$

A further irrational number which was conjectured by Champernowne to be normal in base 10 is formed by the concatenation of the primes:

$$0.235711131719232931374143\ldots.$$

It took until 1946 for A. H. Copeland and the inimitable Paul Erdős to provide the proof and so bring to the mathematical world the Copeland-Erdős Constant.

Finally, Erdős returns with H. Davenport to prove that, if $P(x)$ is a polynomial in x taking positive integer values whenever x is a positive integer, then the number

$$0.P(1)P(2)P(3)\ldots$$

formed by concatenating the base 10 values of the polynomial at $x = 1, 2, 3, \ldots$ is normal in base 10.

---

[7]The reader might also wish to compare the number with $\frac{10}{81} - \frac{3340}{3267}10^{-9}$.

We may as well try to capture Wallis's Proteus of chapter 2 than define a random decimal expansion and when we do have a precise formulation to work with, working with it is currently impractically hard. At present we must accept that all experimental evidence points to the standard irrational constants of mathematics having decimal expansions which are 'random' and that we cannot prove the fact, largely because we cannot define the term.

We can end this chapter on a positive note, though, if we allow ourselves an amusing fancy.

Consider coding the alphabet, the space, the integers from 0 to 9 and standard punctuation symbols in some way (perhaps ASCII, for example) and then take a piece of writing and encode it character by character using this scheme; finally concatenate the encoding to form a single, vast positive integer. Any number Normal in base 10 will have this integer appear infinitely often in its decimal expansion so, for example, Champernowne's Constant contains all versions of the Bible, the complete works of Shakespeare, all of the works in the British Library or, indeed, Borges' Library of Babel; in fact all of human writing, past present and future. Something that cannot be said for rationals!

CHAPTER TEN

One Question, Three Answers

There are certainly people who regard $\sqrt{2}$ as something perfectly obvious but jib at $\sqrt{-1}$. This is because they think they
can visualise the former as something in physical space but not
the latter. Actually $\sqrt{-1}$ is a much simpler concept.

Edward Titchmarsh (1899–1963)

We met Mary Cartwright in chapter 4, challenging prospective
undergraduates to prove that π^2 is irrational. Ted Titchmarsh
was her contemporary at Oxford and was later to hold the Savilian Chair of Geometry some 280 years after John Wallis, and
then for a 'mere' 32 years. Our opening quotation from him may
attract questions about the physical nature of $\sqrt{-1}$ but assuredly
the misconception that $\sqrt{2}$ is an obvious concept while $\sqrt{-1}$ a
profound one had long been widespread. Yet, as we have seen in
chapter 2, without irrational numbers analytic geometry and the
idea of limit, with its consequent implications for calculus, were
in serious difficulty. It is natural now to concentrate on the nineteenth century and we will re-enter it with a quotation from one
of its greatest mathematical minds, Niels Abel, whom we earlier
saw was responsible for showing that there are algebraic numbers
incapable of expression by radicals:

> If you disregard the very simplest cases, there is in all of math-
> ematics not a single infinite series whose sum has been rigor-
> ously determined. In other words, the most important parts of
> mathematics stand without a foundation.

The slippery, intuitive grasp of the nature of real numbers was
insufficient to exercise proper control over them and, more than
anything else, what loosened the grip was the precise definition of

what constitutes an irrational number, for without that the exact nature of arithmetic and of limits and continuity could never be understood.

Preparing the Ground

The early part of the nineteenth century saw varied attempts to give meaning to irrational numbers. The German August Crelle is best remembered for his encouragement of the mathematically talented youth of the day and his journal, *Journal für die reine und angewandte Mathematik* (ever referred to as *J. De Crelle*) trumpeted the early work of a galaxy of talent, including one of the foremost architects of nineteenth-century rigour, Karl Weierstrass, and also Abel; on page 205 we have already alluded to Cantor's work appearing there. Crelle's house was a meeting place for intellectual discussion and we meet one particular minor player in the story of irrational numbers when Abel, in a letter to his mentor, the astronomer Christopher Hansteen, wrote:

> At Crelle's house there used to be a weekly meeting of mathematicians, but he had to suspend it because of a certain Martin Ohm with whom nobody could get along due to his terrible arrogance.

This 'guest from hell' is better remembered as the younger brother of the German physicist Georg Ohm: even more positively, he appears to have been the first to coin the phrase *Golden Ratio* for that most irrational number $\varphi = \frac{1+\sqrt{5}}{2}$, and he made his own attempts to give a definition of irrational numbers. From his 1829 work, *The Spirit of Mathematical Analysis: And Its Relation to a Logical System* he considered the density of the rationals in the following manner.

Since, for any positive integer n, $7n-1 < 7n < 7n+1$ and since $\frac{7}{3} = 7n/3n$ it must be that $(7n - 1)/3n < \frac{7}{3} < (7n + 1)/3n$; the differences can be made arbitrarily small as n becomes arbitrarily large and so he finds the definition of three fractions being *continuous to one another* and thence to a shaky distinction between the rational and irrational in the paragraph:

> Among the fractions which lie continuously to one another, by far the greater number have an infinitely great denominator, and then also a corresponding infinitely great numerator (whereas the fraction itself is by no means infinitely

great, but lies between two other fractions that are near one another, and have finite numerators and denominators). These are called irrational numbers, and whole and broken numbers with finite numerators and denominators are then termed (in contradistinction to the rational) irrational numbers.

And, more promisingly:

The division of decimal fractions, if it were to be accurately completed, would lead to an infinite series, generally termed an "irrational decimal fraction", for which in applications to the comparisons of magnitudes an approximate value may be substituted. In such an irrational decimal fraction the sum of n decimal places continually approaches nearer and nearer to a certain determinate, not infinitely great limit, even when n itself is supposed to be infinitely great; i.e. such an irrational decimal fraction is always what is subsequently termed a convergent infinite series.

And Ohm was not alone. Sir William Rowan Hamilton, the Irish Astronomer Royal, inventor of quaternions, the first person to give a formal construction of the complex numbers and all-round genius, offered a view of the real numbers, not as points on a line but, prosaically, as instances of time. Further, in his two papers read before the Royal Irish Academy in 1833 and 1835 and published as *Algebra as the Science of Pure Time*, he introduced the idea of repeatedly separating the rationals into two classes and defining an irrational number to be the partition generated, but he did not complete the work. Had the Bohemian mathematician, logician, theologian and philosopher Father Bernard Bolzano published his thoughts there would be many areas of intellectual enquiry that would be associated with his name, including an 1835 manuscript presaging one of the accepted forms of the definition of irrational numbers. Bolzano's influence should, though, be compared with that of Augustin-Louis Cauchy (1789–1857). In his vastly influential work *Cours d'analyse* of 1821 he stated that:

An irrational number is the limit of diverse fractions which furnish more and more approximate values of it.

Cauchy was one of the great nineteenth-century rigorists, and gave a precise definition of convergence of a sequence to a limit:

when the successive values attributed to a variable approach a
fixed value indefinitely so as to end by differing from it as little
as is wished, this fixed value is called the limit of all the others.

Symbolically, the sequence $\{x_n\}$ has a limit x if:

For any $\varepsilon > 0$ there exists an N so that, if $n > N$, $|x_n - x| < \varepsilon$

and we are reminded of Eudoxus's *method of exhaustion.* Yet his
observation about irrational numbers cannot be used to define
them: this definition of convergence requires the limit to exist
and if we only have the rational numbers it may not do so.

His lasting contribution to what was to be one of the ways of
defining an irrational number was a second formulation of the
convergence of a sequence:

A necessary and sufficient condition that a sequence $\{x_r\}$ converges to a limit is that the difference between $\{x_n\}$ and $\{x_m\}$
can be made arbitrarily small for all m, n sufficiently large.

It is important to realize that the condition cannot be weak-
ened to the difference between consecutive terms becoming
arbitrarily small implying that the sequence converges: take
the sequence of partial sums of the harmonic series, $\{x_n\}$,
with $x_n = \sum_{r=1}^{n} (1/r)$, which has its consecutive terms differing precisely by the harmonic sequence itself. The harmonic
sequence evidently approaches 0 but the harmonic series is (not
so evidently) famously divergent to infinity as we mentioned on
page 138, which makes its sequence of partial sums diverge to
infinity.

Certainly, the *necessary* part follows from the definition of
convergence since, if the sequence converges:

$$|x_n - x_m| = |(x_n - x) - (x_m - x)| \leqslant |x_n - x| + |x_m - x| < \varepsilon.$$

But it is the *sufficient* part that carries with it what is needed to
define irrational numbers, provided care is taken not to assume
the existence of the limit, as we shall see.

Then came an 1867 publication *Prinzip der Permanenz der for-
malen Gesetze* of Hermann Hankel in which he asked whether
there were other number systems which had essentially the same
rules as the real numbers; it seems to have been the first published reference to the approach of Weierstrass, which we mention
below, but its author evidenced his scepticism with:

Every attempt to treat the irrational numbers formally and without the concept of (geometric) magnitude must lead to the most abstruse and troublesome artificialities, which, even if they can be carried through with complete rigour, as we have every right to doubt, do not have a higher scientific value.

It was the Frenchman Charles Méray (1835–1911) who was the first to publish a coherent theory of irrational numbers, which appeared in 1870 as part of the report[1] of the 1869 congress of the Sociétés Savantes. First but not foremost as, in the words of no less an authority than Henri Poincaré:

Unfortunately, since he used a special language that he created and which repelled most of his readers, his influence was in fact mediocre.

The Three Answers

What, then, is an irrational number? The ground, if still strewn with logical bumps, had been sufficiently prepared for some to attach an arithmetic meaning to real numbers, thereby detaching number from magnitude, and after a wait of two millennia, four publications appeared in Germany in the same year. We may deem 1872 to be the *Year of the Irrational* with the publication of:

- *Die Elemente der Arithmetik* by Ernst Kossak;
- *Die Elemente der Functionenlehre*[2] by Eduard Heine;
- *Ueber die Ausdehnung eines Satzes aus der Theorie der Trigonometrischen Reihen*[3] by Georg Cantor;
- *Stetigkeit und Irrationale Zahlen* by Richard Dedekind.

The first two publications were disseminations of the work of Weierstrass, who had outlined his approach to irrational numbers as the set of infinite decimal expansions in a series of lectures given in Berlin in the 1860s; he was assisted by Georg Cantor. We may move past Kossack's contribution without loss; Weierstrass did so having considered it to be a *disorganized jumble*.

---

[1] Remarques sur la nature des quantités définies par la condition de servir de limites à des variables données (Remarks on the nature of the quantities defined by the condition of use limits on variable data).

[2] *J. De Crelle* 74:172-88.

[3] A paper primarily concerned with Fourier series, which appeared in *Math. Annalen* 5:123-32.

Heine's contribution was based on what he had learned of Weier-strass's work from Cantor and echoed the more general approach of Cantor himself and was to be later further developed by Otto Stolz.

The translated title of Cantor's article is 'On the extension of a theorem of the theory of trigonometric series', which seems removed from the theory of irrational numbers, but in it he found himself considering infinite sets of points in relation to the problem of the convergence of series and, in order to operate rigorously, he then proposed an arithmetic theory of irrational numbers. He embraced Cauchy's ideas of the convergence of a sequence, carefully avoiding circularity and thereby generalized the approach of Weierstrass.

Richard Dedekind (1831–1916) held the chair at the Polytech-nikum in Zürich in 1858, an appointment that brought with it a responsibility to teach introductory differential calculus, the first time that the 27-year-old professor of mathematics had been called upon so to do – and he was unhappy in his endeavours: in his own (translated) words[4]:

> In discussing the notion of the approach of a variable mag-nitude to a fixed limiting value, and especially in proving the theorem that every magnitude which grows continually, but not beyond all limits, must certainly approach a limiting value, I had to resort to geometric evidences. ... But this form of introduction into the differential calculus can make no claim to being scientific, no one can deny. For myself this feeling of dissatisfaction was so overpowering that I made the fixed resolve to keep meditating on the question till I should find a purely arithmetic and perfectly rigorous foundation for the principles of infinitesimal analysis.

Having done so, from the same work, he later commented:

> ... and in this way we arrive at real proofs of theorems, for example $\sqrt{2} \times \sqrt{3} = \sqrt{6}$, which to the best of my knowledge have never been established before.

With commendable temporal precision, he tells us in the same volume that it was November 24, 1858 (a Wednesday) when he

---

[4]Richard Dedekind, 2003, *Continuity and Irrational Numbers*, included in *Essays on the Theory of Numbers* (Dover).

thought of the real line being populated by the ordered rationals separated by gaps, which are the irrational numbers. With
the image of the number line being cut into two pieces by each
gap the concept of the *schnitt*, or cut, came into being. The idea
is reminiscent of the independent approach of Hamilton, but we
have now what is routinely called the *Dedekind cut*. When he realized that others like Heine and Cantor were about to publish their
versions of a rigorous definition of the real numbers he decided
that he too should publish his own ideas.

   More publications soon followed but contributed nothing that
was significantly new; in essence, with the approaches of Cantor
and Dedekind, the problem of irrationality was solved. We shall
take a brief look at what are now the three standard formulations
of the real number system, with the first subsumed by the second, and we may think of the differences in approach as being
arithmetic, analytic and algebraic.

The Weierstrass–Heine Model

The real numbers are defined to be the set of infinite decimal
expansions

$$\mathbb{R} = \{a_0.a_1a_2a_3\cdots :$$
$$a_0 \in \mathbb{Z}, \ a_r \in \{0,1,2,3,4,5,6,7,8,9\}, \ r > 0\},$$

which are interpreted as sum of the infinite series

$$a_0 + \frac{a_1}{10} + \frac{a_2}{10^2} + \frac{a_3}{10^3} + \cdots .$$

Notice that they must be *infinite* decimal expansions, which brings
about a subtlety with, for example, $2.7000\ldots = 2.6999\ldots$. Further, although the infinite series and decimal form are the same
for positive numbers, they are not for numbers less than 0. For
example, since $-2.7 \neq -2 + \frac{7}{10}$ and we must write $-2.7 = -3 + \frac{3}{10}$.
Defining the basic arithmetic operations is an extremely delicate
matter, with each of them performed on pairs of infinitely long
sequences of numbers, with possible carries causing great logis-
tical difficulty. In the end the difficulties can be overcome and
rational numbers appear as those whose expansions are infinitely
repeating or ones which end in an infinite sequence of 9s, leaving

the remainder to be the irrationals. With all of this subtle (and rather tedious) detail omitted we have a first proper model of the real number system. The real numbers are defined to be

$$\mathbb{R} = \text{the set of all infinite decimal expansions.}$$

The Cantor-Heine-Méray Model

If we consider the sequence of rational numbers, continued in the obvious manner, $3, 3.1, 3.14, 3.141, 3.1415, 3.14159, \ldots$, we are well aware that it converges to the irrational number π so that, if the nth term of the sequence is x_n then, using Cauchy's rigorous formulation of limit, given any rational $\varepsilon > 0$ there is a positive integer n so that $|x_n - \pi| < \varepsilon$. We could think of defining π as the limit of this sequence, but we have already mentioned that our argument would be circular: we are using the limit itself to define the limit itself and we need to know how to define an irrational number as the limit of an infinite sequence, without assuming that limit's existence.

We return to the idea of a *Cauchy sequence* (of rational numbers), $\{x_r\}$, as a sequence for which all terms eventually differ by arbitrary small amounts; symbolically:

For any rational $\varepsilon > 0$ there exists a positive integer N so that $|x_r - x_s| < \varepsilon$ for all $r, s > N$.

Now let us consider those infinite decimal expansions in general, focusing on the decimal part. Taking $x = 0.n_1 n_2 n_3 \ldots$ we have the sequence of rational truncations:

$$\{x_i\} = \{0.n_1 n_2 n_3 \ldots n_i\}$$
$$= \{n_1 \times 10^{-1} + n_2 \times 10^{-2} + n_3 \times 10^{-3} + \cdots + n_i \times 10^{-i}\}.$$

Then, for $j > i$,

$$|x_i - x_j| = n_{i+1} \times 10^{-(i+1)} + n_{i+2} \times 10^{-(i+2)}$$
$$+ n_{i+3} \times 10^{-(i+3)} + \cdots + n_j \times 10^{-j} < 10^{-i}.$$

For sufficiently large i (and therefore j) this can be made arbitrarily small and so $\{x_i\}$ is a Cauchy sequence and the Weierstrass approach is thereby embraced.

Evidently, if q is any rational number, $x_r = \{q, q, q, \ldots\}$ is a Cauchy sequence and we reasonably identify the Cauchy sequence with the rational number; every other Cauchy sequence is defined to be its own a real number. Arithmetic is largely easier than the previous model with, for example,

$$0 = \{0, 0, 0, \ldots\} \quad \text{and} \quad 1 = \{1, 1, 1, \ldots\}.$$

And for two real numbers $\alpha = \{x_r\}$, $\beta\{y_r\}$

$$\alpha + \beta = \{x_r + y_r\} \quad \text{and} \quad \alpha \times \beta = \{x_r y_r\},$$
$$-\alpha = \{-x_r\} \quad \text{and} \quad \frac{1}{\alpha} = \left\{\frac{1}{x_r}\right\}.$$

We also have a natural definition of inequality with

$\alpha < \beta$ if and only if there exists rational $\varepsilon > 0$ and there exists a natural number N so that

$$x_n + \varepsilon < y_n \quad \text{whenever } n \geqslant N.$$

Surprisingly, defining equality is a little more subtle, with the subtlety stemming from an intrinsic ambiguity. Take, for example, the two Cauchy sequences:

$$x_r = \{q, q, q, \ldots\} \quad \text{and} \quad y_r = \{0, 0, 0, \ldots, q, q, q, \ldots\},$$

where there are a finite number of 0s: they each represent the rational number q. More generally, take any Cauchy sequence and manufacture a second by changing a finite number of terms of it; both are Cauchy sequences and both converge to the same limit. Alternatively, take any infinite subsequence of a Cauchy sequence and the same is true. We need to deal with this ambiguity and do so using the idea of equivalence:

Two Cauchy sequences $\{x_r\}$ and $\{x_r'\}$ are said to be *equivalent* (written $\{x_r\} \sim \{x_r'\}$) if, for every rational $\varepsilon > 0$, there exists a positive integer N so that $|x_r - x_r'| < \varepsilon$ for $r > N$; that is, eventually their difference approaches 0.

The condition does define an equivalence relation[5] and all we need do is ensure that all equivalent Cauchy sequences converge

---

[5] See Appendix E on page 289.

to the same limit. To see this, suppose that $x_r \xrightarrow{r\to\infty} x$ and that $x'_r \xrightarrow{r\to\infty} x'$, with $\{x_r\} \sim \{x'_r\}$. Then

$$|x' - x| = |(x_r - x) - (x'_r - x') - (x_r - x'_r)|$$
$$< |x_r - x| + |x'_r - x'| + |x_r - x'_r|.$$

Since each of the three terms on the right can be made arbitrarily small for sufficiently large values of r, the discrepancy between the two constants on the left is arbitrarily small, leaving the only option that they are, in fact, the same.

With this, we modify the original statement and define real numbers to be the equivalence classes of Cauchy sequences under this equivalence relation, and this allows the definition of equality to be

$$\alpha = \beta \quad \text{if and only if} \quad \{x_r\} \sim \{y_r\}.$$

As before, there is much to check and not least that we can use any member of any equivalence class to represent the corresponding real number. We will be content with the following:

The Cauchy sequences

$$\{x_r\} = \Big\{ \underbrace{1, 1, 1, \ldots, 1}_{r \text{ terms}} \Big\} \quad \text{and} \quad \{x'_r\} = \Big\{ \underbrace{0.9, 0.99, 0.999, \ldots}_{r \text{ terms}} \Big\}$$

are equivalent since

$$\{x_r - x''_r\} = \Big\{ \underbrace{(1 - 0.9), (1 - 0.99), (1 - 0.999), \ldots}_{r \text{ terms}} \Big\}$$
$$= \Big\{ \underbrace{0.1, 0.01, 0.001, \ldots}_{r \text{ terms}} \Big\}$$

and so $|x_r - x'_r| \xrightarrow{r\to\infty} 0$, which gives us that $0.999\ldots = 1$.

We have, then, a second realization of the real numbers:

$$\mathbb{R} = \text{the set of all equivalence classes of Cauchy sequences.}$$

The Dedekind Model

A Dedekind cut is a pair of non-empty, non-intersecting subsets $L, R \subset \mathbb{Q}$ satisfying

(a) $L \cup R = \mathbb{Q}$;

(b) For all $a \in L$ and $b \in R$, $a < b$;

(c) L contains no largest element.

In consequence, for any rational number q, $L = \{r \in \mathbb{Q}: r < q\}$ and $R = \{r \in \mathbb{Q}: r \geqslant q\}$. It is clear that knowledge of the composition of L automatically reveals the composition of R and so we can represent a cut simply by the letter of the first set; for example,

$$\tfrac{22}{7} = \{r \in \mathbb{Q}: r < \tfrac{22}{7}\}$$

We might think of these as 'closed' cuts; what is of interest are those that are 'open'. Quoting Dedekind:

> In every case in which a cut is given that is not produced by a rational number, we create a new number, an irrational number, which we consider to be completely defined by this cut; we will say that the number corresponds to this cut or that it produces the cut.

Take, for example, $\sqrt{2}$, which is defined to be the cut

$$\sqrt{2} = \{r \in \mathbb{Q}: r \leqslant 0\} \cup \{r \in \mathbb{Q}_+: r^2 < 2\}$$

and a more complicated example,

$$2^{2/3} = \{r \in \mathbb{Q}: r \leqslant 0\} \cup \{r \in \mathbb{Q}_+: r^3 < 4\}.$$

Comparison of two numbers is easy here with the definition of equality of two real numbers α and β

$$\alpha = \beta \quad \text{if they are equal as sets}$$

and inequality $\alpha < \beta$ is defined by set inclusion $\alpha \subset \beta$.

Of course, being rational numbers, we have

$$0 = \{r \in \mathbb{Q}: r < 0\} \quad \text{and} \quad 1 = \{r \in \mathbb{Q}: r < 1\}.$$

Addition is as natural as can be with

$$\alpha + \beta = \{a + b: a \in \alpha, b \in \beta\}$$

and multiplication too for positive numbers with

$$\alpha \times \beta = \{r \in \mathbb{Q}: r \leqslant 0\} \cup \{ab: a \in \alpha, b \in \beta \text{ with } a, b > 0\}.$$

Matters begin to be more subtle when we have to have regard for negative numbers.

Of course, the number $-3 = \{r \in \mathbb{Q}: r < -3\}$, but this must also result from the negation of $3 = \{r \in \mathbb{Q}: r < 3\}$ itself.

To relate 3 to its negative, -3, we should proceed

$$-3 = \{r \in \mathbb{Q}: r < -3\} = \{r \in \mathbb{Q}: -r > 3\},$$

which produces the R section of the cut, making the L section

$$-3 = \{r \in \mathbb{Q}: -r \leqslant 3\},$$

which is not a cut since it has a greatest element. To correct matters we need to eliminate the possibility of equality and do so by defining

$$-\alpha$$
$$= \{r \in \mathbb{Q}: -r > \alpha \text{ and } -r \text{ is not the least element of } \mathbb{Q} - \alpha\}$$
$$= \{r \in \mathbb{Q}: \text{there exists } x > 0 \text{ such that } -r - x > \alpha\}.$$

With this awkwardness set aside we can define general multiplication by

$$\alpha \times \beta = \begin{cases} 0 & \text{if } \alpha = 0 \text{ or } \beta = 0, \\ |\alpha| \times |\beta| & \text{if } \alpha < 0 \text{ and } \beta < 0, \\ -(|\alpha| \times |\beta|) & \text{if } \alpha > 0 \text{ and } \beta < 0 \text{ or } \alpha < 0 \text{ and } \beta > 0, \end{cases}$$

where

$$|\alpha| = \begin{cases} \alpha: & \alpha > 0, \\ -\alpha: & \alpha < 0. \end{cases}$$

Finally, we will mention the multiplicative inverse since it is subject to the same subtlety as its additive counterpart above with

$$\alpha^{-1} = \{r \in \mathbb{Q}: r \leqslant 0\} \cup \{a > 0 \text{ and } 1/a > \alpha \text{ and } 1/a$$
$$\text{is not the least element of } \mathbb{Q} - \alpha\}$$
$$= \{a \in \mathbb{Q}: \text{there exists } b > 0 \text{ such that } -1/a - b \notin \alpha\}$$

under the assumption that $\alpha > 0$. If $\alpha < 0$, $\alpha^{-1} = -(|\alpha|)^{-1}$.

As to Dedekind's theorem that $\sqrt{2} \times \sqrt{3} = \sqrt{6}$, we have

$$\sqrt{2} = \{r \in \mathbb{Q}: r \leqslant 0\} \cup \{r_1 \in \mathbb{Q}_+: r_1^2 < 2\}$$

and

$$\sqrt{3} = \{r \in \mathbb{Q}: r \leqslant 0\} \cup \{r_2 \in \mathbb{Q}_+: r_2^2 < 3\}$$

and, from the definition of product,

$$\sqrt{2} \times \sqrt{3} = \{r \in \mathbb{Q}: r \leqslant 0\} \cup \{r_1 r_2 \in \mathbb{Q}_+: r_1^2 < 2 \text{ and } r_2^2 < 3\}$$
$$= \{r \in \mathbb{Q}: r \leqslant 0\} \cup \{r_1 r_2 \in \mathbb{Q}_+: r_1^2 r_2^2 < 6\}$$
$$= \{r \in \mathbb{Q}: r \leqslant 0\} \cup \{r_1 r_2 \in \mathbb{Q}_+: (r_1 r_2)^2 < 6\}$$
$$= \sqrt{6}.$$

And with this success we will leave matters but with the memory that

$$\mathbb{R} = \text{the set of Dedekind cuts.}$$

A Reconciliation

So, are the real the numbers the set of Cauchy sequences or the set of Dedekind cuts? Of course, they must be both and here we will suggest a bridge which might be used to move from a real number defined in one way to one defined in the other.

First, the Cauchy sequence $\{q, q, q, \ldots\}$ is associated with the Dedekind cut $\{r \in \mathbb{Q}: r < q\}$.

Now, if we take an irrational number $x = \{x_n\}$ we will manufacture its Dedekind cut as follows:

For each rational number q consider the sequence $\{x_n - q\}$. If $\{x_n - q\} \geqslant 0$ for sufficiently large n then $\lim_{n \to \infty} \{x_n - q\} \geqslant 0$ and so $x - q > 0$ and $q < x$ and should be placed in L, otherwise $q > x$ and so $q \in R$.

Conversely:

1. Start with the irrational number defined by the cut (L, R).
2. Choose some rational number $x_L \in L$ and $x_R \in R$.
3. Define $x_0 = x_L$ and $x_1 = x_R$.
4. Set the next term of the sequence to be $a = (x_L + x_R)/2$.
5. If the rational number a lies in L call it our new x_L, otherwise call it our new x_R.
6. Return to step 4.

It is clear that $\{x_n\}$ is a Cauchy sequence converging to the cut; further, it is easily seen that starting with different rational numbers produces equivalent Cauchy sequences.

Abstraction

A long and perceptive look at what we need of our real number system reveals that there are in essence two arithmetic operations of $+$ and $\times$ which behave and combine in acceptable ways, a sense of order between the numbers and a further sense of completeness of them.

These various qualities were abstracted out in 1900 by the German mathematical giant David Hilbert to provide a theoretical model of the real numbers, one that we can use as a measuring stick for any realization of them. We shall look at this abstraction but first the context is set with the earlier work of Paul Tannery with his first abstraction of the rational numbers.

If we allow ourselves to start with Kronecker's *God-given* integers,[6] the rational numbers can be defined as equivalence classes of pairs of them (a, b) with $b \neq 0$ under the equivalence relation[7] $\sim$ where

$$(a, b) \sim (c, d) \quad \text{if and only if } ad = bc.$$

Addition, multiplication and ordering are achieved by

- $(a, b) + (c, d) = (ad + bc, bd)$;
- $(a, b) \times (c, d) = (ac, bd)$;
- $(a, b) < (c, d) \Leftrightarrow ad < bc$.

All manner of things have to be checked but if they are we will find what we recognize as our rational number system, with only a notational change from (a, b) to a/b required to be on completely familiar territory. Furthermore, if we make a tactical jump and assume that the real numbers have been satisfactorily defined, it is just as easy to construct the complex numbers from them in much the same manner:

- $(a, b) + (c, d) = (a + c, b + d)$,
- $(a, b) \times (c, d) = (ac - bd, ad + bc)$,

---

[6]'God made the integers, all else is the work of man' – a famous remark he is said to have made during an after-dinner speech.

[7]See Appendix E on page 289.

with the association $(a, b) = a + ib$. We have evidence, then, of the perceptive nature of Titchmarsh's comment.

With the scene thus set, we will look carefully at what is required of a number system that is rich enough to fulfil the needs of mathematical demands, with those requirements naturally dividing into three categories.

The Arithmetic Axioms

The axioms for arithmetic assert that, when we combine two real numbers using addition $+$ or multiplication $\times$, a real number results and we have:

For addition:

1. There exists an *additive identity* $0 \in \mathbb{R}$, so that $0+a = a+0 = a$ for all $a \in \mathbb{R}$.
2. For each $a \in \mathbb{R}$ there exists an *additive inverse*, denoted by $-a$, so that $a + (-a) = (-a) + a = 0$.
3. The *associative law* holds: that is, $(a + b) + c = a + (b + c)$ for all $a, b, c \in \mathbb{R}$: that is, the expression $a + b + c$ is unambiguous.
4. The *commutative law* holds: that is, $a + b = b + a$ for all $a, b \in \mathbb{R}$.

For multiplication:

1. There exists a *multiplicative identity* $1 \in \mathbb{R}$, so that $1 \times a = a \times 1 = a$ for all $a \in \mathbb{R}$.
2. For each $a(\neq 0) \in \mathbb{R}$ there exists a *multiplicative inverse*, denoted by a^{-1}, so that $a \times a^{-1} = a^{-1} \times a = 1$.
3. The *associative law* holds: that is, $(a \times b) \times c = a \times (b \times c)$ for all $a, b, c \in \mathbb{R}$: that is, the expression $a \times b \times c$ is unambiguous.
4. The *commutative law* holds: that is, $a \times b = b \times a$ for all $a, b \in \mathbb{R}$.

Finally, the *distributive law* ensures that addition and multiplication are compatible:

For all $a, b, c \in \mathbb{R}$, $a \times (b + c) = a \times b + a \times c$, that is, we can multiply out brackets.

With these we have all that is needed for finite arithmetic and from them we can deduce all familiar arithmetic results such as

$0 \times a = 0$, which means that 0 has no multiplicative inverse, and the fact that 'minus times minus is a plus': $(-1) \times (-1) = 1$.

We have not reached the real numbers yet. In general, a set with two operations defined on its elements (called addition and multiplication) that satisfy these axioms is called a *field* and examples of fields abound: in particular, not only do the real numbers form a field but the rational numbers and the complex numbers do also. We must demand more to shed the simpler systems.

The Order Axioms

The order axioms assert that there is a relation $<$ defined on all elements of $\mathbb{R}$ which satisfies the following rules:

1. The *trichotomy law* asserts that exactly one of the relations $a < b, b < a, a = b$ holds between any $a, b \in \mathbb{R}$.
2. The *law of transitivity* asserts that if $a < b$ and $b < c$ then $a < c$.
3. The *law of compatibility with addition* asserts that if $a < b$ then $a + c < b + c$ for all $c \in \mathbb{R}$.
4. The *law of compatibility with multiplication* asserts that if $a < b$ then $a \times c < b \times c$ for all $0 < c \in \mathbb{R}$.

Of course, we have the symbols $>$, $\geqslant$, $\leqslant$ for convenience with their obvious meanings. If we insist on these being attached to the field axioms we finish with an *ordered field* and, although the rational numbers are one such, here we leave the complex numbers behind, since a conclusion from the axioms of order is that, for all $a \neq 0$, $a \times a > 0$.

One major step remains.

The Completeness Axiom

If every nonempty subset of the ordered field with an *upper bound* has a *least upper bound*, the field is called a *complete ordered field*. Let us expand on this. Consider a subset S of our ordered field, then b is an *upper bound* of S if $s \leqslant b$ for all $s \in S$ and b is the *least upper bound* (it can be shown to be unique) if $b < c$ for any other upper bound c of S. It is with this axiom that we leave the rational numbers behind since, for example, the set $S = \{s \in \mathbb{Q} : s \times s < 2\}$ has any number of upper bounds (take 2, for example) but no least upper bound – since $\sqrt{2}$ is irrational. This final structure has the name of a complete ordered field.

What is left? The real numbers. That is, it is possible to prove that there is one and only one complete ordered field: it is this structure that is the abstraction of the real numbers. All properties that we associate with real numbers can be established from these axioms, whether it be that all real numbers can be written as an infinite decimal, that the triangle inequality holds ($|a + b| \leqslant |a| + |b|$) or that the Axiom of Eudoxus from chapter 1 holds: if $a, b > 0$ there is an integer $n > 0$ so that $na > b$. When we suggested that the details associated with Cauchy sequences and Dedekind cuts can be checked, those details are precisely these axioms for a complete ordered field. With this, those two realizations are but different views of the same unique construction and we declare satisfaction that the real number system has now been satisfactorily defined. In the 1898 words of the German mathematician Carl Thomae:

> The formal conception of numbers requires of itself more modest limitations than does the logical conception. It does not ask, what are and what shall the numbers be, but it asks, what does one require of numbers in arithmetic.

This said, this recondite area of mathematical foundations has attracted as much controversy as it has interest and we will leave it with a view of Hilbert's contemporary, the great German mathematical philosopher Gottlob Frege, who would have nothing of the constructions of Hilbert, Weierstrass, Cantor or Dedekind (et al.) and, as late as 1903, wrote:

> This task[8] has never been approached seriously, let alone been solved.

---

[8]Of defining the real numbers.

CHAPTER ELEVEN

Does Irrationality Matter?

> If the ratio of the circumference of a circle to its diameter be
> written to thirty five places of decimals, the result will give the
> whole circumference of the visible universe without an error
> as great as the minutest length visible in the most powerful
> microscope
>
> Simon Newcomb (1835-1909)

The opening quotation is taken from Newcomb's 1882 book
*Logarithmic and Other Mathematical Tables: With Examples of
Their Use and Hints on the Art of Computation.* Microscopes have
become a great deal more powerful than those available to him
and so have telescopes and in consequence we would need to
increase that number of decimal places considerably for his statement now to hold true. Yet true it would then be and, moreover, he
declared that 'for most practical applications' five-figure accuracy
is quite sufficient. For the man whose shared logarithmic tables
used at the Nautical Almanac Office alerted him to what is now
known as Benford's Law, $\pi = 3.14159$.

Practical Matters

From the very lengthy lists of formulae involving π and e let us
select four from each:

- The Cauchy distribution has a density function
  $\varphi(x) = 1/\pi(1 + x^2)$.
- The probability that two randomly chosen integers are
  coprime is $6/\pi^2$.
- The period of a simple pendulum is $T = 2\pi\sqrt{l/g}$.

- Einstein's field equations are
  $R_{ik} - g_{ik}R/2 + \Lambda g_{ik} = (8\pi G/c^4)T_{ik}$.
- The logarithmic spiral has polar equation $r = e^{a\theta}$.
- The catenery has Cartesian equation $y = (e^x + e^{-x})/2$.
- The damped harmonic oscillator is governed by
  the equation $x = e^{-t}\cos(\omega t + \alpha)$.
- The Poisson distribution has density function
  $\varphi(k) = e^{-\lambda}\lambda^k/k!$.

And then there is the density function of the normal distribution, $\varphi(x) = (1/\sqrt{2\pi})e^{-x^2/2}$, which conveniently involves all three of our standard irrational triplet of π, e and $\sqrt{2}$.

The theoretical derivation of every one of these formulae, and all besides, leads to the appearance of some fundamental constant(s): π, e and $\sqrt{2}$ are three such. They appear as exact numbers and so it is as well that we have symbols for them, but it is quite irrelevant that they are irrational. For all practical applications any awkward constant that appears in a formula can be approximated as necessary so that the theoretical formula is utilized to help with a problem of the real world, with a well-known mathematical 'joke' putting matters into a light-hearted perspective:

> What is π?
> A mathematician: π is the ratio of the circumference of a circle to its diameter.
> A computer programmer: π is 3.141592653589 in double precision.
> A physicist: π is 3.14159 plus or minus 0.000005.
> An engineer: π is about 22/7.
> A nutritionist: pie is a healthy and delicious dessert.

In chapter 2 we mentioned the skill demonstrated by the Hindus and Arabs in manipulating surds but their ability to approximate was equally impressive. The astronomer and mathematician Al-Khwārizmī, whom we mentioned there, used the three approximations to π: $\frac{22}{7}$, $\sqrt{10}$ and $\frac{62832}{20000}$, where he described the first as *an approximate value*, the second *as used by geometricians* and the third *as used by astronomers*. The values were not new with all of them, and many more besides, originating with the Hindus and reaching the Arabs through translations of works; this mysterious last value, impressively accurate to four decimal places, seems to have originated with the Hindu polymath Aryabhata 1 (476–550 C.E.) and is found in his work *Aryabhatiyam*:

> 8 times 100 increased by 4 and 62 of 1000s is an approximate circumference of a circle diameter 20,000[1]

which makes $\pi \sim \frac{8(100+4)+62\times1000}{20000} = 3.1416$. Quite how he came about the approximation is a mystery.

Throughout millennia, with the Greeks and their predecessors, long before the Hindus and Arabs and long after them, methods to approximate irrational numbers have tested the ingenuity of the scientist, with each new equation bringing with it its own requirements for sensible application. Let us take a contemporary example: if we refer to the Universal Constants page of the website of the National Institute of Standards and Technology[2] we find a table, one entry of which is the *permeability of free space* μ_0, otherwise known as the magnetic constant, which has a defined value of exactly $4\pi \times 10^7$ N A^{-2}. The value then ascribed to the constant for purposes of calculation is $12.566370614\ldots \times 10^7$, which means that, in this case, $3.1415926533 \leqslant \pi < 3.1415926536$.

Now move to $\sqrt{2}$ and its involvement in the well-tempered musical scale. This twentieth-century division of the musical scale has the octave defined as the interval between a note and a second of twice its frequency. This frequency interval is divided into 12 subintervals (semitones) between (say) C and the c above with the frequency of c double that of C. The scheme is coded in musical terms as follows:

$$A, A\#, B, C, C\#, D, D\#, E, F, F\#, G, G\#.$$

The well-tempered scale is defined to have the frequency of the next A to the right of middle C at exactly 435 Hz, which means that the frequencies of the subsequent notes in the octave are

$$435 \times (1, 2^{1/12}, 2^{2/12}, 2^{3/12}, 2^{4/12}, 2^{5/12},$$
$$2^{6/12}, 2^{7/12}, 2^{8/12}, 2^{9/12}, 2^{10/12}, 2^{11/12}, 2).$$

---

[1]Victor J. Katz (ed.), 2007, *The Mathematics of Egypt, Mesopotamia, China, India and Islam* (Princeton University Press).

[2]physics.nist.gov/cgi-bin/cuu/Category?view=gif&
Universal.x=117&Universal.y=10.

And we have a collection of eleven irrational multiples with the standard accuracy of two decimal places giving the values

$$(435, 460.87, 488.27, 517.31, 548.07, 580.67, 615.18,$$
$$651.76, 690.52, 731.58, 775.08, 821.17, 870).$$

With this accepted standard of useful accuracy, if our approximation to $2^{1/12}$ is written as T the most testing requirement is that $821.165 < T^{11} \times 435 < 821.175$ and this makes

$$1.05946243 < T < 1.05946326.$$

Since $2^{1/12} = 1.059463094\ldots$ we can safely use the approximation $T = 1.059463$ and six decimal places is sufficient for purpose.

And finally let us move to the Golden Ratio, $\varphi = (-1 + \sqrt{5})/2$. It is commonly held that rectangles of dimension $1 \times \varphi$ are aesthetically pleasing and that this harmony has been exploited by architects and artists from antiquity to the present day. Yet, it is hard to imagine visiting the artist or the architect with a calculating task more challenging than $\varphi \sim 1.6$ and $1/\varphi \sim 0.6$: the Parthenon will not fall as a result or the Mona Lisa look less beautiful. More demanding of precision is the use of φ in speaker cabinet design. It is agreed by experts that a simple cuboidal speaker will have its undesirable internal standing waves significantly diminished if its dimensions are in the ratio $1/\varphi : 1 : \varphi$. Let us see what an expert[3] advises on the optimal dimensions for a speaker of volume 4 cubic feet:

> First, take the desired enclosure volume and convert it into a more precise ('working') unit of measurement – in your case inches. 1 cubic foot equals 1728 cubic inches; therefore 4 cubic feet equals 6912 cubic inches.
>
> Next, take the cubed root of the total volume, 6912 cubic inches. The result becomes the basic working dimension of the enclosure. For a 6912 cubic inch enclosure this will be 19.0488 inches.
>
> Now take 19.0488 inches and multiply it by the quantity $(-1 + \sqrt{5})/2$, or 0.6180339887, and you will obtain one of the

---

[3] answers.yahoo.com/question/index?qid=20080406124528AAfNJia.

other 'optimum' dimensions, which should be 11.7728 inches
for the smallest internal dimension for your enclosure.

Finally, take the initial 19.0488 inch result and multiply it by
the quantity $(-1 + \sqrt{5})/2$, or 1.6180339887, and you will obtain
the remaining optimum dimension, which should be 30.8216
inches for the largest internal dimension for your enclosure.

If you wish to verify the validity of your calculations
simply multiply the three results together and you should
obtain almost exactly the original enclosure volume. Thus,
(11.7728 in.)(19.0488 in.)(30.8216 in.) = 6911.9993 in.^{3} approximately. Again, I have to emphasize that it is very important
to remember that all of the dimensions above are INTERNAL
enclosure dimensions.

And so we have the optimal dimensions, each to four decimal
places. But then, the advice finishes with the statement:

Also please note that if you prefer you can safely round each
of the derived dimensions to the nearest half-inch (0.50 in.)
without incurring any detrimental consequences.

So, in the end the optimum dimensions of the loudspeaker box
are $12 \times 19 \times 31$ inches.

With such a final tolerance the accuracy required is determined
by the two double inequalities:

$$11.5 \leqslant \sqrt[3]{6912} \times \frac{1}{\varphi} < 12.5 \quad \text{and} \quad 30.5 \leqslant \sqrt[3]{6912} \times \varphi < 31.5.$$

To manipulate these to bounds for φ we will use the full accuracy
that a standard calculator gives for the cube root of the volume
to see that

$$1.5239 < \varphi < 1.6564 \quad \text{and} \quad 1.6011 < \varphi < 1.6536.$$

The second condition is, of course, the stronger and so we may
take the value of φ to be anything in the interval $1.6011 < \varphi <
1.6536$. Removing the awkward volume and replacing it with a
more convenient one of, say, 1000 units a recalculation yields the
interval $1.55 < \varphi < 1.65$. So, it seems that $\varphi \sim 1.61$ would do
nicely. When the author asked a friend who is a sound engineer
what value would be taken as φ in his own calculations the immediate response was 1.618, which seems to make a great deal of
sense.

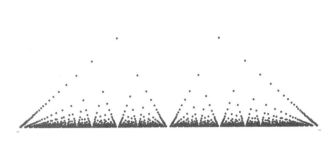

Figure 11.1.

Irrationality is quite irrelevant, then: now we will consider the other case, where it is crucial.

Theoretical Matters

With the definition of the function $\chi(x)$,

$$\chi(x) = \begin{cases} 1: & x \text{ is rational,} \\ 0: & x \text{ is irrational,} \end{cases}$$

it is hardly surprising that the rationality or otherwise of a number is of crucial importance. What may be surprising is the observation that the function given on page 8 is an analytic representation of it, with

$$\chi(x) = \lim_{m \to \infty} \lim_{n \to \infty} \cos^{2n}(m!\pi x) = \begin{cases} 1: & x \text{ is rational,} \\ 0: & x \text{ is irrational.} \end{cases}$$

As such it is an indicator function for the (ir)rational numbers and a shocking example of an infinite sequence of functions the (double) limit of which is everywhere discontinuous.

It seems that Dirichlet was the first to consider such a pathology and it seems also that it was the German Carl Thomae who amended it to

$$f(x) = \begin{cases} 1/q & \text{if } x = p/q \text{ is a rational number in lowest terms,} \\ 0 & \text{if } x \text{ is irrational.} \end{cases}$$

This *Thomae* (*Popcorn, Star over Babylon*, etc.) function, shown in figure 11.1, also has an obvious dependency on (ir)rationality

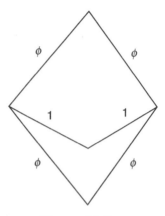

Figure 11.2.

and a rather less obvious consequence of this dependency is that it is continuous at all irrational numbers and discontinuous at all rational numbers.

The informal demonstration of the fact is pleasant and uses an observation from chapter 6: if we want to approximate a real number arbitrarily closely by rational numbers, the denominators of those irrational numbers must grow arbitrarily large. With this reminder, suppose that x is irrational and so $f(x) = 0$. If x_1 is arbitrarily close to x then x_1 is either irrational, in which case $f(x_1) = 0$, or rational, in which case $f(x_1) \sim 0$ since its denominator will be arbitrarily large. These combine to $f(x)$ being continuous at irrational values for x. On the other hand, suppose that x is rational, then again the arbitrarily close point x_1 is either irrational and so $f(x_1) = 0$ or rational and so $f(x_1) \sim 0$ and these are not arbitrarily close to $f(x) = 1/q$.

The dependency on irrationality is more subtle with the version of the *Penrose Tiles* consisting of the dart and the kite, as shown in figure 11.2. These form an aperiodic tiling of the plane only because the Golden Ratio is irrational, otherwise the tiling would be periodic.

Now consider three results from the study of dynamical systems, which are often framed in terms of a circle:

Take a circle C of circumference 1 unit, choose any real number α and repeatedly shift each point (say anticlockwise) on the circle by α, measured as a distance along the circumference. With this, every point will orbit the circle indefinitely, but in a manner depending on whether α is rational or irrational:

If $\alpha = p/q$ in lowest terms is rational the orbit of every point will be closed, repeating itself after q iterations.

If α is irrational the whole process is far more complex, with these results referring to the orbit of any point on the circle:

1. The orbits 'exhaust the circle' in that, if we take any point on the circle there is an infinite subsequence of the orbit which converges to the point.
2. For any arc of length l on the circle, the asymptotic fraction of points in the orbit which are contained in the arc is precisely l.
3. If we truncate the orbit to a finite set of points, these points divide the circle into intervals of at most three different lengths and if there are three different lengths the largest is the sum of the other two.

In fact, that first result was foretold by Nicole Oresme, whom we first met in chapter 2 attempting to use irrational numbers to dismantle the uncomfortable idea of the Perfect Year. In Proposition II.4 of his work, Tractatus de commensurabilitate vel incommensurabilitate motuum celi,[4] he made the following comment regarding two bodies moving in a circle with uniform but incommensurable velocities:

No sector of a circle is so small that two such mobiles could not conjunct in it at some future time and could not have conjuncted in it sometime (in the past).

We could continue with such results but will choose to move to three cases where irrationality is crucial. In the first two they appear in a manner reminiscent of the role of complex numbers when they are used to prove results about the reals: they appear, solve the problem and disappear; rather like a mathematical fairy godmother. In the third case there is a geometric result of no little interest.

Waclaw Sierpiński's rather cryptic response to a question posed in 1957 by Hugo Steinhaus was this: the point $(\sqrt{2}, \frac{1}{3})$ has different distances from all points of the plane lattice.[5]

---

[4]On the commensurability or incommensurability of the heavenly motions
[5]The points in the plane with both coordinates integers.

First, let us see why.

Suppose that, with integer coordinates, there are two points (a_1, b_1) and (a_2, b_2) that are equidistant from $(\sqrt{2}, \frac{1}{3})$. Then

$$(a_1 - \sqrt{2})^2 + (b_1 - \tfrac{1}{3})^2 = (a_2 - \sqrt{2})^2 + (b_2 - \tfrac{1}{3})^2,$$
$$a_1^2 + b_1^2 - a_2^2 - b_2^2 - \tfrac{2}{3}(b_1 - b_2) = 2\sqrt{2}(a_1 - a_2).$$

The left side is rational and the right irrational unless $a_1 = a_2$ and this makes the left side 0. So,

$$a_1^2 + b_1^2 - a_2^2 - b_2^2 - \tfrac{2}{3}(b_1 - b_2) = 0$$

and, using $a_1 = a_2$,

$$b_1^2 - b_2^2 - \tfrac{2}{3}(b_1 - b_2) = (b_1 - b_2)((b_1 + b_2) - \tfrac{2}{3}) = 0.$$

Since the two points are assumed distinct and since the first coordinates are already proved to be the same, $b_1 \neq b_2$, and therefore

$$(b_1 + b_2) - \tfrac{2}{3} = 0,$$

which is impossible since b_1, b_2 are integers.

Steinhaus's question initially seems remote from Sierpiński's answer, since it was: For any positive integer n, does there exist a circle which contains precisely n points having integer coordinates?

The emphasis is with the condition that there should be *precisely* n points and the answer is *yes* and is found in the following pleasant argument.

We have seen that Cantor had long before proved that the integers $\mathbb{Z}$ and the plane lattice points $\mathbb{Z} \times \mathbb{Z}$ are each countable. We have already noted that this means that the sets can be listed, so let us list the points of the plane lattice as

$$\mathbb{Z} \times \mathbb{Z} = \{L_1, L_2, L_3, \ldots\},$$

where the ordering is by increasing distance from the point $(\sqrt{2}, \frac{1}{3})$. Then,

$$|x - (\sqrt{2}, \tfrac{1}{3})| < |L_{n+1} - (\sqrt{2}, \tfrac{1}{3})|$$

defines a circular disc centre $(\sqrt{2}, \frac{1}{3})$ and radius $|L_{n+1} - (\sqrt{2}, \frac{1}{3})|$, which contains precisely the points $L_1, L_2, L_3, \ldots, L_n$.

Our second example requires a more lengthy discussion. Take the positive integers and divide them into two infinite, non-empty, non-overlapping subsets. These subsets may have structure or not: if one is the set of prime numbers and the other the set of composite numbers there is structure but no pattern, with one the set of even numbers and the other odd numbers there is both and with one a random selection there is neither. We will look at a case where there is clear structure which sometimes gives rise to a pattern and at other times not.

In the 1877 book *The Theory of Sound* authored by the eminent mathematical physicist John Strutt, 3rd Baron Rayleigh (Lord Rayleigh), we find an observation made by him in the study of the harmonics of a vibrating string which can naturally be framed in purely mathematical terms in terms of irrational numbers. Independently rediscovered by the American mathematician Samuel Beatty,[6] it appeared in the problems column of the *American Mathematical Monthly* in the words:

3173. Proposed by Samuel Beatty, University of Toronto.

If x is a positive irrational number and y its reciprocal, prove that the sequences

$$(1 + x), \quad 2(1 + x), \quad 3(1 + x), \quad \ldots$$
$$(1 + y), \quad 2(1 + y), \quad 3(1 + y), \quad \ldots$$

contain one and only one number between each pair of consecutive positive integers.

Solutions were published in the journal in March of the following year[7] and the problem was resurrected in 1959 as Problem A6 of that year's Putnam competition[8] in a slightly revised form:

Show that, if x and y are positive irrationals such that $1/x + 1/y = 1$ then the sequence $\lfloor x \rfloor, \lfloor 2x \rfloor, \ldots \lfloor nx \rfloor, \ldots$ and $\lfloor y \rfloor, \lfloor 2y \rfloor, \ldots \lfloor ny \rfloor, \ldots$ together include every positive integer exactly once.[9]

And it is this form that we shall study the phenomenon.

---

[6] *American Mathematical Monthly*, March 1926, Problem 3173, 33(3):159.

[7] *American Mathematical Monthly*, March 1927, Solution 3177, 34(3):159-60.

[8] William Lowell Putnam Mathematical Competition: Problems and Solutions 1938-1964, 1980, *Math. Ass. Am.*, pp. 513-14.

[9] Where, once more, $\lfloor \cdot \rfloor$ is the Floor function.

First notice that a rational value for x will not do: write $x = p/q$, where $p > q$ (since $x > 1$), then a common term will be reached when $\lfloor (p/q)n \rfloor = \lfloor (p/(p-q))m \rfloor$ and with $n = q$ and $m = p-q$ or multiples thereof the common integer p and its multiples are generated.

Before we consider the matter, let us restate it as:

> Let α be a positive irrational number, define the irrational number β by $1/\alpha + 1/\beta = 1$ and the two sets A and B by
>
> $$A = \{\lfloor n\alpha \rfloor : n = 1, 2, 3, \dots \} \quad \text{and} \quad B = \{\lfloor n\beta \rfloor : n = 1, 2, 3, \dots \}.$$
>
> Then:
> $$A \cap B = \varnothing, \qquad A \cup B = \mathbb{N}.$$

If this is the case we should note that the two sequences A and B are termed *complementary*.

We know that irrationality is necessary for the result to hold, before we show that it is sufficient let us develop a feel for the process by considering these four examples.

Take $\alpha = \sqrt{2}$, then $\beta = \frac{\sqrt{2}}{\sqrt{2}-1} = 2 + \sqrt{2}$ and we have

$$\begin{aligned}
A = \{ &\lfloor 1\sqrt{2} \rfloor, \lfloor 2\sqrt{2} \rfloor, \lfloor 3\sqrt{2} \rfloor, \lfloor 4\sqrt{2} \rfloor, \lfloor 5\sqrt{2} \rfloor, \lfloor 6\sqrt{2} \rfloor, \lfloor 7\sqrt{2} \rfloor, \\
&\lfloor 8\sqrt{2} \rfloor, \lfloor 9\sqrt{2} \rfloor, \lfloor 10\sqrt{2} \rfloor, \lfloor 11\sqrt{2} \rfloor, \lfloor 12\sqrt{2} \rfloor, \lfloor 13\sqrt{2} \rfloor, \\
&\lfloor 14\sqrt{2} \rfloor, \lfloor 15\sqrt{2} \rfloor, \lfloor 16\sqrt{2} \rfloor, \lfloor 17\sqrt{2} \rfloor, \dots \},
\end{aligned}$$

$$\begin{aligned}
B = \{ &\lfloor 1(2 + \sqrt{2}) \rfloor, \lfloor 2(2 + \sqrt{2}) \rfloor, \lfloor 3(2 + \sqrt{2}) \rfloor, \lfloor 4(2 + \sqrt{2}) \rfloor, \\
&\lfloor 5(2 + \sqrt{2}) \rfloor, \lfloor 6(2 + \sqrt{2}) \rfloor, \lfloor 7(2 + \sqrt{2}) \rfloor, \lfloor 8(2 + \sqrt{2}) \rfloor, \\
&\lfloor 9(2 + \sqrt{2}) \rfloor, \lfloor 10(2 + \sqrt{2}) \rfloor, \lfloor 11(2 + \sqrt{2}) \rfloor, \lfloor 12(2 + \sqrt{2}) \rfloor, \\
&\lfloor 13(2 + \sqrt{2}) \rfloor, \lfloor 14(2 + \sqrt{2}) \rfloor, \lfloor 15(2 + \sqrt{2}) \rfloor, \\
&\lfloor 16(2 + \sqrt{2}) \rfloor, \lfloor 17(2 + \sqrt{2}) \rfloor, \dots \},
\end{aligned}$$

which simplify to

$$A = \{1, 2, 4, 5, 7, 8, 9, 11, 12, 14, 15, 16, 18, 19, 21, 22, 24, \dots \},$$

$$\begin{aligned}
B = \{ &3, 6, 10, 13, 17, 20, 23, 27, 30, \\
&34, 37, 40, 44, 47, 51, 54, 58, \dots \}.
\end{aligned}$$

Similarly, if $\alpha = \pi$, the sequences are

$$A = \{3, 6, 9, 12, 15, 18, 21, 25, 28, 31, 34,$$
$$37, 40, 43, 47, 50, 53, 56, 59, 62, \dots\},$$
$$B = \{1, 2, 4, 5, 7, 8, 10, 11, 13, 14, 16, 17,$$
$$19, 20, 22, 23, 24, 26, 27, 29, \dots\}.$$

And if $\alpha = e$ they are

$$A = \{2, 5, 8, 10, 13, 16, 19, 21, 24, 27, 29,$$
$$32, 35, 38, 40, 43, 46, 48, 51, 54, \dots\},$$
$$B = \{1, 3, 4, 6, 7, 9, 11, 12, 14, 15, 17, 18,$$
$$20, 22, 23, 25, 26, 28, 30, 31, \dots\}.$$

Lastly, and provocatively, if $\alpha = \gamma + 1^{10}$

$$A = \{1, 3, 4, 6, 7, 9, 11, 12, 14, 15, 17, 18,$$
$$20, 22, 23, 25, 26, 28, 29, 31, \dots\},$$
$$B = \{2, 5, 8, 10, 13, 16, 19, 21, 24, 27, 30,$$
$$32, 35, 38, 40, 43, 46, 49, 51, 54, \dots\}.$$

With these suggesting the truth of the result, we shall go about proving it.

First we prove that $A \cap B = \varnothing$ and to do this let us assume otherwise and presume that the two sequences have an element in common. If this is so, there must exist positive integers m and n so that $\lfloor n\alpha \rfloor = \lfloor m\beta \rfloor = N$ and this means that the two inequalities

$$N < n\alpha < N + 1 \quad \text{and} \quad N < m\beta < N + 1$$

must simultaneously hold: note the crucial point that the inequalities are strict since α and β are irrational. Rearranging these gives

$$\frac{n}{N+1} < \frac{1}{\alpha} < \frac{n}{N} \quad \text{and} \quad \frac{m}{N+1} < \frac{1}{\beta} < \frac{m}{N}.$$

---

[10]Euler's Constant, $\gamma = 0.57721\dots$, the irrationality of which we have remarked remains an open question. Also, we need the number used to be greater than 1.

Adding results in

$$\frac{n}{N+1} + \frac{m}{N+1} < \frac{1}{\alpha} + \frac{1}{\beta} < \frac{n}{N} + \frac{m}{N}.$$

And using the defining condition

$$\frac{n+m}{N+1} < 1 < \frac{n+m}{N}.$$

And rearranging this double inequality results in

$$N < n + m < N + 1,$$

which tells us that there is an integer $m + n$ lying strictly between the integers N and $N + 1$ - which is impossible. It must be that $A \cap B = \varnothing$.

Now consider the two finite sequences, the biggest element of one being N and of the other an integer less than N:

$$A_N = \{\lfloor n\alpha \rfloor : \lfloor n\alpha \rfloor \leqslant N;\ n = 1, 2, 3, \ldots\},$$
$$B_N = \{\lfloor n\beta \rfloor : \lfloor n\beta \rfloor < N;\ n = 1, 2, 3, \ldots\}.$$

For example, with the $\sqrt{2}$ case above and with $N = 12$,

$$A_{12} = \{1, 2, 4, 5, 7, 8, 9, 11, 12\} \quad \text{and} \quad B_{12} = \{3, 6, 10\}.$$

A little thought reveals that the number of elements in each sequence is calculated as follows:

$$|A_{12}| = 9 = \left\lfloor \frac{12+1}{\sqrt{2}} \right\rfloor \quad \text{and} \quad |B_{12}| = 3 = \left\lfloor \frac{12+1}{2+\sqrt{2}} \right\rfloor.$$

And, in general,

$$|A_N| = \left\lfloor \frac{N+1}{\alpha} \right\rfloor \quad \text{and} \quad |B_N| = \left\lfloor \frac{N+1}{\beta} \right\rfloor.$$

So, the total number of elements in the two lists is

$$|A_N| + |B_N| = \left\lfloor \frac{N+1}{\alpha} \right\rfloor + \left\lfloor \frac{N+1}{\beta} \right\rfloor$$
$$= \left\lfloor \frac{N+1}{\alpha} \right\rfloor + \left\lfloor (N+1)\left(1 - \frac{1}{\alpha}\right) \right\rfloor$$
$$= \left\lfloor \frac{N+1}{\alpha} \right\rfloor + \left\lfloor (N+1) - \frac{N+1}{\alpha} \right\rfloor.$$

Now we need two properties of the Floor and Ceiling functions:

$$\lfloor M + \theta \rfloor = M + \lfloor \theta \rfloor \quad \text{and} \quad \lfloor -\theta \rfloor = -\lceil \theta \rceil$$

for M an integer and θ arbitrary. Using these, we have

$$|A_N| + |B_N| = \left\lfloor \frac{N+1}{\alpha} \right\rfloor + (N+1) + \left\lfloor -\frac{N+1}{\alpha} \right\rfloor$$
$$= \left\lfloor \frac{N+1}{\alpha} \right\rfloor + (N+1) - \left\lceil \frac{N+1}{\alpha} \right\rceil.$$

Since $(N+1)/\alpha$ cannot be an integer, $\lfloor (N+1)/\alpha \rfloor = \lceil (N+1)/\alpha \rceil - 1$ and this means that

$$|A_N| + |B_N| = \left\lceil \frac{N+1}{\alpha} \right\rceil - 1 + (N+1) - \left\lceil \frac{N+1}{\alpha} \right\rceil = N.$$

And we are done: the two sequences combine to the sequence of positive integers $\{1, 2, 3, \ldots, N\}$ for arbitrary N.

Now look back at our four examples. The alert reader will have noticed gaps in the latter portion of $A \cup B$ since we are constrained by the inevitable stopping point in the lists A and B: much more than this, we are condemned to use a rational approximation to the numbers to generate the lists. In short, we can prove the result but never realize it.

Let us look finally at a more general construction which separates the positive integers.

In 1954 two Canadian mathematicians Leo Moser and Joachim (Jim) Lambek established the following result about complementary sequences[11]:

> Let $f(n)$ be a non-decreasing function which maps the natural numbers into itself and define a second function $f^*(n) =$ the maximum k for which $f(k) < n$ for each natural number n. Now define two more functions $F(n) = f(n) + n$ and $G(n) = f^*(n) + n$, then the sequences generated by $F(n)$ and $G(n)$ will be complementary.

At first the result is a little slippery to grasp so let us take a particular case, with $f(n) = 2n$. The process generates the following

---

[11]J. Lambek and L. Moser, 1954, Inverse and complementary sequences of natural numbers, *American Mathematical Monthly* 61:454–58.

table:

n	1	2	3	4	5	6	7	...
$f(n)$	2	4	6	8	10	12	14	...
$f^*(n)$	0	0	1	1	2	2	3	...
$F(n)$	3	6	9	12	15	18	21	...
$G(n)$	1	2	4	5	7	8	10	...

We can see that the final two rows of the table suggest a partition the natural numbers, with our two sets defined to be

$$A = \{3, 6, 9, 12, 15, 18, 21, \ldots\}$$

and

$$B = \{1, 2, 4, 5, 7, 8, 10, \ldots\}.$$

The multiples of 3 and the non-multiples of 3.

Moreover, we have a simple formula for $F(n) = f(n) + n = 2n + n = 3n$, which is hardly a surprise. What is much more exiting is that we can also find a compact formula for the B sequence of non-multiples of 3; that is, for the function $G(n) = f^*(n) + n$.

If we look more carefully at the definition of f^*, to find its value for any n we need to find the maximum k for which the equation $2k < n$ holds, that is, $k < n/2$. This suggests the Floor function with maximum $k = \lfloor n/2 \rfloor$, but the inequality is strict and if n were to be even $\lfloor n/2 \rfloor = n/2$. To combat this and to cope with both cases simultaneously, for odd n the fractional part of $n/2$ is precisely $\frac{1}{2}$ and so subtracting (say) $\frac{1}{4}$ will not alter matters and will bring the value of the Floor function to its required level. That is, the maximum $k = \lfloor n/2 - \frac{1}{4} \rfloor$ and so $f^*(n) = \lfloor n/2 - \frac{1}{4} \rfloor$, which makes

$$G(n) = f^*(n) + n = \lfloor n/2 - \tfrac{1}{4} \rfloor + n$$

and we have a succinct formula for sequence B and so for all non-multiples of 3. Try it! And afterwards the reader might like to use the same technique[12] to show that the nth non-square natural number can be written

$$G(n) = \lfloor \sqrt{n} + \tfrac{1}{2} \rfloor + n$$

---

[12]This time, start with $F(n) = n^2$.

and, without hint, the nth non-triangular number

$$G(n) = \lfloor \sqrt{2n} + \tfrac{1}{2} \rfloor + n$$

and then move to other possibilities.

Our main interest is that we can frame our earlier result involving irrational numbers in these terms.

Define

$$f(n) = \lfloor n\alpha \rfloor - n \quad \text{and so } F(n) = \lfloor n\alpha \rfloor.$$

By its definition, $f^*(n) = $ the largest m so that $f(m) < n$. That is, the largest m so that $\lfloor m\alpha \rfloor - m < n$ and so $\lfloor m\alpha - m \rfloor = \lfloor m(\alpha - 1) \rfloor < n$. This in turn means that $\lfloor m(\alpha - 1) \rfloor = n - 1$ and this means that $n - 1 < m(\alpha - 1) < n$. Finally, it must therefore be that m is the largest integer so that $m < n/(\alpha - 1)$, which makes $m = \lfloor n/(\alpha - 1) \rfloor$. All of this means that $f^*(n) = \lfloor n/(\alpha - 1) \rfloor$ and this makes

$$G(n) = \left\lfloor \frac{n}{\alpha - 1} \right\rfloor + n = \left\lfloor \frac{n}{\alpha - 1} + n \right\rfloor = \left\lfloor n \frac{1 + \alpha - 1}{\alpha - 1} \right\rfloor$$

$$= \left\lfloor n \frac{\alpha}{\alpha - 1} \right\rfloor$$

$$= \lfloor n\beta \rfloor,$$

where $\alpha/(\alpha - 1) = \beta$, or $1/\alpha + 1/\beta = 1$. The reader will identify why the irrationality is important.

Can we separate the positive integers into three or more disjoint, infinite sets using irrational numbers? We cannot, as Aviezri S. Fraenkel has provided a delightful proof of the following result[13]:

Let $m > 1$, $\alpha_1, \ldots, \alpha_m$ be positive. Then $\{\lfloor n\alpha_i \rfloor : i = 1, \ldots, m; n = 1, 2, \ldots\}$ is a complementary system if and only if

1. $m = 2$,

2. $1/\alpha_1 + 1/\alpha_2 = 1$,

3. α_1 is irrational.

[13] Aviezri S. Fraenkel, 1977, Complementary systems of integers, *American Mathematical Monthly* 84:114–15.

There are, though, the likes of

$$\{\{\lfloor\varphi\lfloor n\varphi\rfloor\rfloor\},\{\lfloor\varphi\lfloor n\varphi^2\rfloor\rfloor\},\{\lfloor n\varphi^2\rfloor\}: n = 1,2,3,\dots\}$$

or

$$\{\lfloor\varphi\lfloor n\varphi\rfloor\rfloor: n = 1,2,3,\dots\}$$
$$= \{1,4,6,9,12,14,17,19,22,25,27,$$
$$30,33,35,38,40,43,46,48,51,\dots\},$$
$$\{\lfloor\varphi\lfloor n\varphi^2\rfloor\rfloor: n = 1,2,3,\dots\}$$
$$= \{3,8,11,16,21,24,29,32,37,42,45,$$
$$50,55,58,63,66,71,76,79,84,\dots\},$$
$$\{\lfloor n\varphi^2\rfloor: n = 1,2,3,\dots\}$$
$$= \{2,5,7,10,13,15,18,20,23,26,28,$$
$$31,34,36,39,41,44,47,49,52,\dots\},$$

which has been shown to be complementary by the Norwegian mathematician Thoralf Skolem, who (among other things) has also produced the two-dimensional equivalent, which does not necessarily involve irrational numbers, but when it does so:

If α_1 is irrational then $\{S(\alpha_1,\beta_1), S(\alpha_2,\beta_2)\}$ form a partition if and only if

$$\frac{1}{\alpha_1} + \frac{1}{\alpha_2} = 1 \quad \text{and} \quad \frac{\beta_1}{\alpha_1} + \frac{\beta_2}{\alpha_2} \text{ is an integer,}$$

where

$$S(\alpha_i,\beta_i) = \{\lfloor\alpha_i n + \beta_i\rfloor: n = 1,2,3,\dots; i = 1,2,3\dots\}.$$

We will leave the matter of the importance of irrationality on what might be regarded as a negative note: an example wherein it is crucial for a number *not* to be irrational.

Finally, we shall consider the problem of the dissection of a square into squares (squaring the square), which seems to be one of surprisingly recent vintage. The English puzzle genius Henry Dudeney appears to have been the first to hint at the idea with his *Lady Isobel's Casket* problem, which appeared in the *Strand Magazine* of January 1902. In fact the problem asked for a dissection of a 20×20 square into squares and a single $10 \times \frac{1}{4}$ rectangle

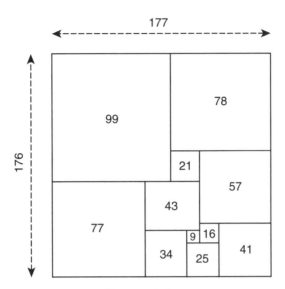

Figure 11.3.

but the idea quickly caught the imagination and results connected with the problem rapidly appeared and are now legion. Our interest is with a contribution of Max Dehn, who, in 1903, proved that a *rectangle* can be squared if and only if the ratio of its sides is a rational number. Figure 11.3 shows an example of such a squaring and the result a final opportunity for proof.

First we show that, if the ratio of the sides of the rectangle is rational, the rectangle can be squared. Actually, it can be tiled as a bathroom floor with squares of side 1 unit. If the sides of the rectangle $a \times b$ are both integers the tiling by unit squares is obvious; if $a/b = p/q$ is rational the rectangle is similar to the one with sides $qa \times qb = pb \times qb$ and can be tiled by squares of side b.

Now suppose that the rectangle is capable of being tiled by squares. We set up a coordinate system with origin the lower left corner of the square and sides the x and y axes and, if it happens that both coordinates of every corner of every square are integers, the side of every square would be an integer and so the sides of the rectangle must be integers.

Suppose, then, that this is not and recall this version of Dirichlet's Approximation Theorem from chapter 6:

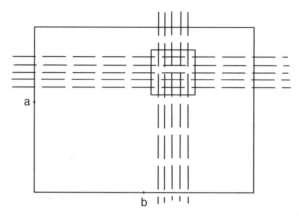

Figure 11.4.

Suppose that α is a real number, then for any positive integer n there exist positive integers p and q, with $q \leqslant n$, such that $|q\alpha - p| < 1/(n + 1)$.

That is, a suitable integer multiple of any real number can be made arbitrarily close to some integer.

Suppose, though, that we have two real numbers α and β. Then the result asserts that

$$|q\alpha - p| < \frac{1}{n + 1} \quad \text{and} \quad |q_1\beta - p_1| < \frac{1}{n + 1}$$

for suitable integers, but what is not clear is that it is possible to choose $q_1 = q$ for some q; that is, the inequalities can be simultaneously satisfied. Generalizing this to an arbitrary finite number of real numbers brings us to the Dirichlet Simultaneous Approximation Theorem, which tells us that this is possible. We choose not to prove this difficult result but certainly we shall now use it.

We refer to figure 11.4 throughout. Although not all coordinates of all vertices of all squares are integers we invoke the theorem and so multiply each of them by the same integer to create a similar rectangle tessellated by squares of near-integer coordinate vertices. Now draw horizontal lines half a unit apart throughout the rectangle and consider the total length L_h of these lines within the rectangle. First, each component of L_h has length a and there are an integral number n_a of them, therefore $L_h = a \times n_a$; similarly, constructing vertical lines results in $L_v = b \times n_b$.

Second, we add up the component lengths of L_h by counting the contribution from each square of the tessellation. If the lengths of the sides of the squares are written s_i then $L_h = \sum_i s_i \times n_{s_i}$; again we repeat for the vertical lines, and this is where the near integral coordinates comes in. Since the vertices can be made arbitrarily close to integers, no matter what the location of the square in the rectangle, the vertical lines will cross each square exactly the same number of times as the horizontal lines, therefore $L_v = \sum_i s_i \times n_{s_i}$. This means that $L_h = L_v$ and so $a \times n_a = b \times n_b$, which means that $a/b = n_b/n_b$ is rational.

And with this we are done: our account of irrationality has eventually reached its final page. It has been a tale of long-term struggle, desperate optimism, grudging acceptance, inspired insight and no little controversy. Our twenty-first-century understanding is at once firm and far-reaching, as it is beset with unanswered questions. It is our hope that among this book's readership there will be one (or more) whose fate it is to carry matters further.

The Spiral of Theodorus

We presented the Spiral of Theodorus on page 7. In this appendix we pursue the idea a little further for those readers whose interest has been aroused by the construction. First, the reader may wish to be aware that an alternative name for the spiral is *Quadratwurzelschnecke*, so coined in 1980 by the late Austrian mathematician Edmund Hlawka, who studied a number of its properties.

We reproduce the original diagram below and seek a formula for the coordinates of the non-right-angled vertices of the triangles, for which the use of complex numbers is convenient.

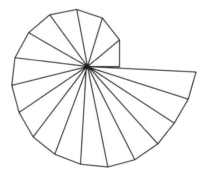

We begin with the vertex $z_1 = 1 + i$ and generate z_{n+1} from z_n by complex addition:

$$z_{n+1} = z_n + \text{ the appropriate complex number of}$$
$$\text{modulus 1 at right angles to } z_n.$$

We recall that multiplying a complex number by i rotates it 90° anticlockwise and that dividing a complex number by its modulus ensures that the result has modulus 1. Thus we generate the

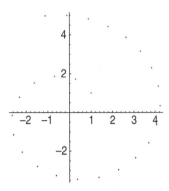

Figure A.1.

spiral as

$$z_1 = 1 + i,$$

$$z_{n+1} = z_n + \frac{z_n}{|z_n|}i, \quad n = 1, 2, 3, \ldots .$$

Since we know that $|z_n| = \sqrt{n+1}$, we can rewrite the generating formula as

$$z_{n+1} = z_n + \frac{z_n}{\sqrt{n+1}}i = \left(1 + \frac{i}{\sqrt{n+1}}\right)z_n$$

and chase this down to

$$z_{n+1} = \left(1 + \frac{i}{\sqrt{n+1}}\right)z_n = \left(1 + \frac{i}{\sqrt{n+1}}\right)\left(1 + \frac{i}{\sqrt{n}}\right)z_{n-1}$$

$$= \left(1 + \frac{i}{\sqrt{n+1}}\right)\left(1 + \frac{i}{\sqrt{n}}\right)\left(1 + \frac{i}{\sqrt{n-1}}\right)z_{n-2}$$

$$= \left(1 + \frac{i}{\sqrt{n+1}}\right)\left(1 + \frac{i}{\sqrt{n}}\right)\left(1 + \frac{i}{\sqrt{n-1}}\right)\cdots(1 + i)$$

$$= \prod_{k=1}^{n+1}\left(1 + \frac{i}{\sqrt{k}}\right).$$

We have, then, the vertices given by

$$z_n = \prod_{k=1}^{n}\left(1 + \frac{i}{\sqrt{k}}\right), \quad n = 1, 2, 3, \ldots .$$

Figure A.1 shows the vertices, isolated from the triangles, with the process continued for a few more terms to emphasize the spiral nature of the construction: it is difficult to resist the temptation to join the dots and so form a continuous curve. In doing so we must, of course, abandon the interpretation of points on the curve as the vertices of the triangles in all but the original discrete cases; how is this best achieved though?

There is a particularly distinguished precedent to guide us: the extrapolation of the discrete factorial function $n! = n \times (n-1) \times (n-2) \times \cdots \times 3 \times 2 \times 1$ to the Gamma Function $\Gamma(x)$. A plot of $n!$ for $n = 1, 2, 3, \ldots$ assuredly suggests to the eye a smooth curve which joins the points and it was the inimitable Leonhard Euler who decided the issue in 1729/30 to leave us with a function equal to $n!$ at positive integers and having meaning at all other numbers, other than the negative integers: the reader unacquainted with the details may wish to pursue the matter further. Our difficulty here lies with the expression for the nth vertex being given as a product of n terms, for which n must necessarily be a positive integer. The device we use to deal with this is not dissimilar to that used by Euler: remove the finite product and replace it with a convergent infinite one. So,

$$
\begin{aligned}
z_n &= \prod_{k=1}^{n} \left(1 + \frac{i}{\sqrt{k}}\right) \\
&= \frac{\left\{\prod_{k=1}^{n} \left(1 + \frac{i}{\sqrt{k}}\right)\right\} \times \left(1 + \frac{i}{\sqrt{n+1}}\right) \times \left(1 + \frac{i}{\sqrt{n+2}}\right) \times \cdots}{\left(1 + \frac{i}{\sqrt{n+1}}\right) \times \left(1 + \frac{i}{\sqrt{n+2}}\right) \times \cdots} \\
&= \prod_{k=1}^{\infty} \left\{ \left(1 + \frac{i}{\sqrt{k}}\right) \Big/ \left(1 + \frac{i}{\sqrt{n+k}}\right) \right\}.
\end{aligned}
$$

Using this we can replace the positive integer variable n by the continuous variable α and so consider the complex-valued function of the real variable α

$$
T(\alpha) = \prod_{k=1}^{\infty} \left\{ \left(1 + \frac{i}{\sqrt{k}}\right) \Big/ \left(1 + \frac{i}{\sqrt{\alpha + k}}\right) \right\}.
$$

The $k = 1$ term in the denominator provides us the restriction $\alpha > -1$ but we can better investigate the function if we convert the product to a sum using logarithms, and then use calculus. So,

$$\ln[T(\alpha)] = \ln\left[\prod_{k=1}^{\infty}\left(1 + \frac{i}{\sqrt{k}}\right) \Big/ \left(1 + \frac{i}{\sqrt{\alpha + k}}\right)\right]$$

$$= \sum_{k=1}^{\infty} \ln\left[\left(1 + \frac{i}{\sqrt{k}}\right) \Big/ \left(1 + \frac{i}{\sqrt{\alpha + k}}\right)\right]$$

$$= \sum_{k=1}^{\infty} \left[\ln\left(1 + \frac{i}{\sqrt{k}}\right) - \ln\left(1 + \frac{i}{\sqrt{\alpha + k}}\right)\right].$$

Now differentiate with respect to α and then rearrange the resulting complex number into its real and imaginary parts:

$$\frac{d}{d\alpha} \ln[T(\alpha)]$$

$$= \frac{T'(\alpha)}{T(\alpha)} = \frac{d}{d\alpha}\left[\sum_{k=1}^{\infty} \ln\left(1 + \frac{i}{\sqrt{k}}\right) - \ln\left(1 + \frac{i}{\sqrt{\alpha + k}}\right)\right]$$

$$= 0 + \sum_{k=1}^{\infty} \frac{i}{2}\left(\frac{1}{(\alpha + k)\sqrt{\alpha + k}}\right) \Big/ \left(1 + \frac{i}{\sqrt{\alpha + k}}\right)$$

$$= \sum_{k=1}^{\infty} \frac{i}{2} \frac{1}{(\alpha + k)\sqrt{\alpha + k} + i(\alpha + k)}$$

$$= \frac{i}{2} \sum_{k=1}^{\infty} \frac{1}{(\alpha + k)(\sqrt{\alpha + k} + i)}$$

$$= \frac{i}{2} \sum_{k=1}^{\infty} \frac{\sqrt{\alpha + k} - i}{(\alpha + k)(\alpha + k + 1)}$$

$$= \frac{1}{2} \sum_{k=1}^{\infty} \frac{1}{(\alpha + k)(\alpha + k + 1)} + \frac{i}{2} \sum_{k=1}^{\infty} \frac{\sqrt{\alpha + k}}{(\alpha + k)(\alpha + k + 1)}$$

$$= \frac{1}{2} \sum_{k=1}^{\infty} \frac{1}{(\alpha + k)(\alpha + k + 1)} + \frac{i}{2} \sum_{k=1}^{\infty} \frac{1}{\sqrt{\alpha + k}(\alpha + k + 1)}$$

$$= \frac{1}{2} \sum_{k=1}^{\infty} \left(\frac{1}{\alpha + k} - \frac{1}{\alpha + k + 1}\right) + \frac{i}{2} \sum_{k=1}^{\infty} \frac{1}{(\alpha + k)^{3/2} + (\alpha + k)^{1/2}}$$

$$= \frac{1}{2} \frac{1}{\alpha + 1} + \frac{i}{2} \sum_{k=1}^{\infty} \frac{1}{(\alpha + k)^{3/2} + (\alpha + k)^{1/2}}$$

with the first series cancelling term-by-term to $1/(\alpha + 1)$.

We have, then,

$$\frac{T'(\alpha)}{T(\alpha)} = \frac{1}{2}\frac{1}{\alpha+1} + \frac{i}{2}\sum_{k=1}^{\infty}\frac{1}{(\alpha+k)^{3/2}+(\alpha+k)^{1/2}}. \qquad (*)$$

Now that we have this convenient expression we can integrate back to find another expression for $T(\alpha)$, with

$$\int_0^{\alpha}\frac{T'(\alpha)}{T(\alpha)}\,d\alpha$$

$$= \frac{1}{2}\int_0^{\alpha}\frac{1}{\alpha+1}\,d\alpha + \frac{i}{2}\int_0^{\alpha}\sum_{k=1}^{\infty}\frac{1}{(\alpha+k)^{3/2}+(\alpha+k)^{1/2}}\,d\alpha,$$

$$[\ln[T(\alpha)]]_0^{\alpha}$$

$$= \tfrac{1}{2}[\ln(\alpha+1)]_0^{\alpha} + \frac{i}{2}\int_0^{\alpha}\sum_{k=1}^{\infty}\frac{1}{(\alpha+k)^{3/2}+(\alpha+k)^{1/2}}\,d\alpha,$$

and with $T(0)=1$ we have

$$\ln T(\alpha) = \tfrac{1}{2}\ln(\alpha+1) + \frac{i}{2}\int_0^{\alpha}\sum_{k=1}^{\infty}\frac{1}{(\alpha+k)^{3/2}+(\alpha+k)^{1/2}}\,d\alpha.$$

All of which makes

$$T(\alpha) = \sqrt{\alpha+1}e^{i\theta},$$

where θ is the nasty expression involving the integral and the infinite sum, with which we will not concern ourselves too much.

Certainly, this form reveals that $|T(\alpha)| = \sqrt{\alpha+1}$, as we would hope, and we can develop the asymptotic polar equation of the spiral in the following manner: we have that $r = \sqrt{\alpha+1}$ and it can be shown that asymptotically $\theta \sim 2\sqrt{\alpha+1} + K$ (see below). This makes the asymptotic form of the polar equation $r \sim \tfrac{1}{2}\theta - \tfrac{1}{2}k$, which is of the form $r = a + b\theta$, the defining equation of the Archimedean spiral: the Theodorus spiral asymptotically approaches an Archimedean spiral. We must admit, then, that our front cover flatters to deceive in that the Nautilus shell is a logarithmic spiral, whose form is $r = ae^{b\theta}$!

Figure A.2 shows a plot of $T(\alpha)$, set in the complex plane. Notice how the behaviour has been extrapolated before the initial point of $1+i$, with the curve crossing the horizontal at 1, where $\alpha = 0$, and spiralling in towards the origin as $\alpha \to -1$ from above.

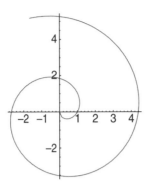

Figure A.2.

Philip J. Davis and others have made a study of this construction and it is here[1] that the above derivation for θ can be found. Davis singled out the slope of the curve as it crosses the horizontal at 1 as a 'fundamental world constant': we will see how it can be evaluated.

First, since $T(\alpha)$ is a complex-valued function of the real variable α, if we write $T(\alpha) = u(\alpha) + iv(\alpha)$, it is standard theory that $T'(\alpha) = u'(\alpha) + iv'(\alpha)$, which makes the gradient of the curve $v'(\alpha)/u'(\alpha)$.

Now, reverting back to equation $(*)$ above and substituting $\alpha = 0$, we have

$$\frac{T'(0)}{T(0)} = \frac{T'(0)}{1} = T'(0) = \frac{1}{2} + \frac{i}{2} \sum_{k=1}^{\infty} \frac{1}{k^{3/2} + k^{1/2}},$$

which tells us that the gradient of the curve as it crosses the horizontal is

$$T = \frac{1}{2} \sum_{k=1}^{\infty} \frac{1}{k^{3/2} + k^{1/2}} \bigg/ \frac{1}{2} = \sum_{k=1}^{\infty} \frac{1}{k^{3/2} + k^{1/2}} = \sum_{k=1}^{\infty} \frac{1}{\sqrt{k}(k+1)},$$

so defining Davis's *Theodorus Constant*,[2] which grudgingly reveals its decimal expansion as

$$T = 1.86002507922119030718069591571714332466652412152345\ldots,$$

and which might well be irrational!

---

[1] Philip J. Davis, 1993, *Spirals from Theodorus to Chaos* (A. K. Peters).

[2] Not to be confused with the commonly held Theodorus Constant that is $\sqrt{3}$.

Rational Parameterizations of the Circle

In chapter 2 we offered the rational parameterization

$$x = \frac{t^2 - 4t - 1}{1 + t^2}, \qquad y = \frac{2 - 2t - 2t^2}{1 + t^2}$$

of the circle $x^2 + y^2 = 5$ and here we disclose its derivation.

First, a reminder of one definition and the statement of another:

- A *rational point* in the plane is one for which both coordinates are rational numbers.
- A *rational line* in the plane is one whose equation can be written as $ax + by + c = 0$, where a, b, c are rational numbers.

We can set up a one-to-one correspondence between the rational points on a circle and the rational points on a rational line in the following way, referring to figure B.1:

- Draw the circle.
- Given there is one, pick any rational point P on the circle to act as a fixed origin.
- Draw a convenient rational line L.
- For every other rational point Q on the circle draw the line PQR, where R is the point of intersection of the line with L.
- The one-to-one correspondence is Q ↔ R.

To establish the one-to-one correspondence, for any rational point Q on the circle, since P is a rational point, the line PQ is a rational line. It is a trivial fact that if two rational lines intersect, they do so in a rational point and this ensures that R, which is the intersection of PQ and L, is a rational point. Conversely, take any rational point R on L and draw the rational line PR. The intersection of this line and our circle will yield a quadratic equation (in x, say, and so of the form $ax^2 + bx + c = 0$) with rational coefficients

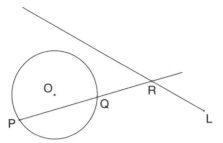

Figure B.1.

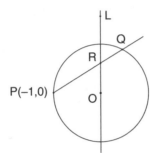

Figure B.2.

and with one of its roots the (rational) x coordinate of P. Since the sum of the two roots $= -b/a$, which is rational, the other root must be rational and we have the x coordinate and therefore y coordinate of the rational point Q on the circle. The one-to-one correspondence is thereby established. For the intersection of the two lines to be assured we must ensure that they are not parallel and for convenience we will take L to be the line perpendicular to OP and passing through O; also, the point P itself is associated with 'the point at infinity' of L.

With the general idea dealt with, we will first see how the method generates the expected algebraic parameterization of the standard unit circle $x^2 + y^2 = 1$.

Take $P = (-1, 0)$ and L the y-axis, as shown in figure B.2. Our line PQR has equation $y = t(1 + x)$, making $R = (0, t)$ and the x coordinate of Q given by the equation

$$1 - x^2 = [t(1 + x)]^2 = t^2(1 + x)^2$$

cancelling by the known factor of $1 + x$ results in $x = (1 - t^2)/(1 + t^2)$ and substituting back into the y equation results in $y =$

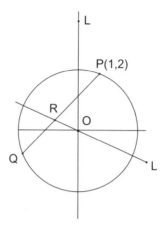

Figure B.3.

$2t/(1+t^2)$ and so we have the standard rational parameterization

$$x = \frac{1 - t^2}{1 + t^2}, \qquad y = \frac{2t}{1 + t^2}.$$

For $x^2 + y^2 = 5$, referring to figure B.3, we will take $P = (1, 2)$, then L has equation $y = -\frac{1}{2}x$. Our line PQR has equation $y - 2 = t(x - 1)$ and this makes

$$R = \left(\frac{2t - 4}{2t + 1}, \frac{2 - t}{2t + 1} \right)$$

and the x coordinate of Q given by the equation $5 - x^2 = [t(x - 1) + 2]^2$, which becomes the quadratic equation

$$(1 + t^2)x^2 + 2t(2 - t)x + (t^2 - 4t - 1) = 0.$$

We know that $x = 1$ is a root and that the sum of the two roots is $-(2t(2 - t))/(1 + t^2)$ and therefore the x coordinate of $Q = -(2t(2 - t))/(1 + t^2) - 1 = (t^2 - 4t - 1)/(1 + t^2)$ and substitution gives its y coordinate as $y = (2 - 2t - 2t^2)/(1 + t^2)$.

So, if there is a single rational point P on the circle, the procedure generates the full infinity of them but fails with the likes of $x^2 + y^2 = 3$ since P simply does not exist. The method applies to conic sections in general.

Two Properties of Continued Fractions

In chapter 3, Lambert's proof of the irrationality of π required two particular properties of continued fractions, which we establish here and, for convenience, restate.

For the continued fraction

$$y = b_0 + \cfrac{a_1}{b_1 + \cfrac{a_2}{b_2 + \cfrac{a_3}{b_3 + \cfrac{a_4}{b_4 + \cdots}}}}.$$

1. If $\{\lambda_1, \lambda_2, \lambda_3, \ldots\}$ is an infinite sequence of non-zero numbers, the continued fraction

$$b_0 + \cfrac{\lambda_1 a_1}{\lambda_1 b_1 + \cfrac{\lambda_1 \lambda_2 a_2}{\lambda_2 b_2 + \cfrac{\lambda_2 \lambda_3 a_3}{\lambda_3 b_3 + \cfrac{\lambda_3 \lambda_4 a_4}{\lambda_4 b_4 + \cdots}}}}$$

has the same convergents as y and, if there is convergence, converges to y.

2. Now suppose that $b_0 = 0$.

If $|a_i| < |b_i|$ for all $i \geqslant 1$, then:

(a) $|y| \leqslant 1$;

(b) writing

$$y_n = \cfrac{a_n}{b_n + \cfrac{a_{n+1}}{b_{n+1} + \cfrac{a_{n+2}}{b_{n+2} + \cfrac{a_{n+3}}{b_{n+3} + \cdots}}}}$$

if, for some n and above, it is never the case that $|y_n| = 1$, then y is irrational.

Both results are susceptible to proof by induction, which we will undertake, but in an informal manner. Unfortunately the formalization of each has something of an unwelcoming shape and we leave it to the reader to manufacture the mould, if this is desired.

Proof of 1

We will simply list the first three cases, which expose the pattern.

Stage 1:

$$b_0 + \frac{a_1}{b_1} = b_0 + \frac{\lambda_1 a_1}{\lambda_1 b_1}.$$

Stage 2:

$$b_0 + \cfrac{\lambda_1 a_1}{\lambda_1 b_1 + \cfrac{\lambda_1 \lambda_2 a_2}{\lambda_2 b_2}} = b_0 + \cfrac{\lambda_1 a_1}{\lambda_1 \left(b_1 + \cfrac{a_2}{b_2}\right)} = b_0 + \cfrac{a_1}{b_1 + \cfrac{a_2}{b_2}}.$$

Stage 3:

$$b_0 + \cfrac{\lambda_1 a_1}{\lambda_1 b_1 + \cfrac{\lambda_1 \lambda_2 a_2}{\lambda_2 b_2 + \cfrac{\lambda_2 \lambda_3 a_3}{\lambda_3 b_3}}} = b_0 + \cfrac{\lambda_1 a_1}{\lambda_1 b_1 + \cfrac{\lambda_1 \lambda_2 a_2}{\lambda_2 \left(b_2 + \cfrac{a_3}{b_3}\right)}}$$

$$= b_0 + \cfrac{\lambda_1 a_1}{\lambda_1 \left(b_1 + \cfrac{a_2}{b_2 + a_3/b_3}\right)}$$

$$= b_0 + \cfrac{a_1}{b_1 + \cfrac{a_2}{b_2 + \cfrac{a_3}{b_3}}}.$$

We hope that, with these, the mystery of the result is laid bare.

Proof of 2

(a) Since

$$|a_{i+1}| < |b_{i+1}|, \qquad \frac{|a_{i+1}|}{|b_{i+1}|} = \left| \frac{a_{i+1}}{b_{i+1}} \right| < 1$$

and so

$$-1 < \frac{a_{i+1}}{b_{i+1}} < 1,$$

which means that

$$b_i - 1 < b_i + \frac{a_{i+1}}{b_{i+1}} < b_i + 1.$$

From this we have, with both sides being positive,

$$b_i + \frac{a_{i+1}}{b_{i+1}} > b_i - 1 \quad \text{and so} \quad \left| b_i + \frac{a_{i+1}}{b_{i+1}} \right| > |b_i - 1|.$$

Using the alternative form of the Triangle Inequality, $|\alpha - \beta| \geqslant |\alpha| - |\beta|$, we then have

$$\left| b_i + \frac{a_{i+1}}{b_{i+1}} \right| > |b_i - 1| \geqslant |b_i| - 1.$$

Now, the a_i and b_i are integers and so it must be that, with $|a_i| < |b_i|$, $|b_i| - 1 \geqslant |a_i|$ and this means that

$$\left| b_i + \frac{a_{i+1}}{b_{i+1}} \right| > |a_i|$$

and so

$$\left| \frac{a_i}{b_i + a_{i+1}/b_{i+1}} \right| < 1.$$

We continue the same argument, working towards the start of the continued fraction.

We have

$$\left| \frac{a_i}{b_i + a_{i+1}/b_{i+1}} \right| < 1$$

so

$$-1 < \frac{a_i}{b_i + a_{i+1}/b_{i+1}} < 1$$

so

$$b_{i-1} - 1 < b_{i-1} + \frac{a_i}{b_i + a_{i+1}/b_{i+1}} < b_{i-1} + 1$$

and

$$\left| b_{i-1} + \frac{a_i}{b_i + a_{i+1}/b_{i+1}} \right| > |b_{i-1} - 1| > |b_{i-1}| - 1 \geqslant |a_{i-1}|,$$

which means

$$\left| \frac{a_{i-1}}{b_{i-1} + \dfrac{a_i}{b_i + \dfrac{a_{i+1}}{b_{i+1}}}} \right| < 1.$$

Continue the process to the start of the continued fraction and we have that the modulus of its expansion to its $i+1$th term is always less than 1 for all i: this must mean that, in the limit, $|y| \leqslant 1$.

(b) Suppose that y is rational and so

$$y = \cfrac{a_1}{b_1 + \cfrac{a_2}{b_2 + \cfrac{a_3}{b_3 + \cfrac{a_4}{b_4 + \cdots}}}} = \frac{p_1}{p_0}.$$

It must be that $|p_1/p_0| \leqslant 1$ and so $|p_1| \leqslant |p_0|$. Now write $y = p_1/p_0 = a_1/(b_1 + y_2)$. Then y_2 must be the rational number

$$y_2 = \frac{a_1 p_0 - b_1 p_1}{p_1} = \frac{p_2}{p_1}$$

and $|y_2| \leqslant 1$ implies that $|p_2| \leqslant |p_1|$. Continue the process indefinitely until we reach n so that $|y_n| < 1$ and therefore $|p_{n+1}| < |p_n|$ and we generate the decreasing infinite sequence of positive integers:

$$|p_0| \geqslant |p_1| \geqslant |p_2| \geqslant \cdots \geqslant |p_n| > |p_{n+1}| > \cdots ,$$

which is clearly impossible. The only reconciliation is that y is irrational.

Finding the Tomb of Roger Apéry

Paris boasts three great cemeteries near its centre: to the north, Montmartre; to the south, Montparnasse and to the east the greatest of them all, in reputation as well as size, Père Lachaise. Located in the 20th arrondissement on one of the city's hills, its name is derived from that of the former owner of the land on which it is sited, who was also confessor to Louis XIV: Père François de La Chaise (1624–1709). In 1804 Napoleon Bonaparte ordered the land to be bought by the city for use as a cemetery, which was later to be extended to its current 119 acres, fully justifying its Parisian epithet: *la cité des morts*; the city of the dead. The reader is advised not to underestimate the complexity of the place. It is estimated that, over the years, one million people have been interred there and it remains an active cemetery for 'ordinary' people as well as for the famous: a list of some of this city's permanent inhabitants is genuinely impressive.[1] The (free) pamphlet provided by the cemetery authorities lists the names and resting places of about 160 of the great and the good and the virtual tour[2] another such list, overlapping but not equal to that on paper: common to both and attractive to the mathematical eye are Gaspard Monge and François Arago, but in neither is the name Roger Apéry. He rests with his parents, with the ashes of all three sealed in the same small tomb set in a wall of the cemetery's Columbarium – a tomb that is far from easy to find. Should readers of this book go to Paris and should fancy have them visit this place and that tomb, we here add a short guide which may circumvent a long search for a small plaque on a concealed wall of a large Columbarium in a vast cemetery.

---

[1] en.wikipedia.org/wiki/P%C3%A8re_Lachaise.

[2] www.pere-lachaise.com/perelachaise.php?lang=en.

The cemetery has three Metro stops in its vicinity (one of which is called Père Lachaise) and a number of bus stops, and it has all of five entrances. We suggest that it is best to seek access via the main entrance, which is named appropriately *Port Principale* and located at the conjunction of *Boulevard Menilmontant* and *Rue du Repos*. Alighting at the metro stop of *Père Lachaise* would then have the reader avoid the corner entrance of *Port des Amandiers* and walk with the cemetery wall on the left to *Port Principale*; the closer Metro stop of *Phillippe-Auguste* requires ignoring the entrance of *Port du Repose* and walking with the cemetery wall to the right to *Port Principale*. The third Metro stop of *Gambetta* is at the rear of the cemetery and significantly closer to our quarry than the other two, yet this convenience should be weighed against the experience of entering this remarkable place along its striking main avenue.

There is an efficiently concealed building within the cemetery wherein the pamphlet can be collected: from *Avenue Principale* turn first right and immediately left and it lies to the right. Whatever diversions are brought about by the paper guide or by idle curiosity, eventually the Columbarium must be found and, with its size and height, it is one of the few easy quarries; it occupies the whole of Division 87 and can be found by taking a route which is a continuation of *Avenue Principale* and leads through the centre of the cemetery, passing close to the chapel at about half way and on and up to the main entrance of the Columbarium itself.

The Columbarium is an open structure in the shape of a cuboid, with the crematorium at its centre. The roof and most of the sides of the cuboid are intentionally missing, with the interior of the residual (two-storey) four corners comprising the walls into which are set the tombs of the cremated – or some of them. These tombs are individually numbered in ascending anticlockwise order from the right of the main entrance; this ensures that the corner to its left contains those with the highest numbers, and these stop at around 7,400: the Apéry tomb is numbered 7,971.

To the initiate it is not obvious that space in the walls has long ago been exhausted and that the authorities have extended occupancy by digging underneath the Columbarium; with this knowledge it is not particularly difficult to locate a subterranean two-storey extension housing thousands more tombs; unfortunately, none of them is tomb number 7,971. It transpires that the first

Figure D.1.

extension – and the one containing the Apéry resting place – was constructed with its only entrance outside the Columbarium in the outside of the wall of its left, rear corner. So, to this corner, which lies at the intersection of the *Avenue des Combattants Étrangers morts pour la France* and the *Avenue Circulaire* and there will be found a small staircase leading down to the corner of an L-shaped room; in the end wall of the left arm, low and to the right is tomb number 7,971. Figure D.1 displays the fruits of the search.

As a bonus a famous, perhaps the most visited and certainly the most kissed[3] tomb in this most visited cemetery, lies in the main burial ground outside and nearby; it is that of Oscar Wilde. But, before leaving the Columbarium, the reader may rise to the challenge of finding tomb number 16,258; it is that of Maria Callas, but hosts an empty urn.

---

[3]In 2011 the tomb was cleaned of its lipstick and a protective glass barrier placed in front of it.

Equivalence Relations

In four places throughout the book we have made mention of an *equivalence relation* on a set. The defining properties of the construction were briefly alluded to in context but here we will look at matters in a little more detail.

First, an equivalence relation may be thought of as a generalization of 'equals'. The universal symbolism $x = y^1$ is replaced by something like $x \sim y$ (x is related to y), although there is no standardization here. With equality it is evident that the two symbols either side of the equals sign are interchangeable, and so it is possible to substitute one for the other in any manipulation; the same is true in its own way with the tilde.

We must start with a set, if not of numbers then of any other well-defined objects, mathematical or not: in this explanation we shall restrict ourselves to the real numbers and subsets of them.

A partition of a set is a collection of non-overlapping subsets which together exhaust the elements of the set: equality partitions the (necessarily) distinct elements into subsets each containing a single element, the tilde partitions into subsets within each of which the elements, although not equal, may be considered in context indistinguishable from one another. Any element in any partition is, for the purposes necessary, representative of all elements in that partition.

How can a relationship defined between the elements of a set bring about a partition of that set? Let us consider some examples.

We will choose the set of integers and define a relationship between them by saying that

$$x \sim y \quad \text{if } x - y \text{ is divisible by 2.}$$

---

[1]First recorded in *The Whetstone of Witte*, the 1557 publication of the Welshman Robert Recorde.

It is clear that this divides the integers into the evens and the odds; an obvious partition.

Alternatively,

$$x \sim y \quad \text{if } x^2 = y^2$$

partitions the integers as $\{\{0\}, \{-1, 1\}, \{-2, 2\}, \{-3, 3\}, \dots\}$.

Finally, consider the set of real numbers with the relation defined as

$$x \sim y \quad \text{if } x - y \text{ is an integer}$$

The resulting partition of the real numbers consists of sets of numbers the decimal parts of which are equal; for example, $\pi, \pi \pm 1, \pi \pm 2, \dots$.

Of course, not all relationships partition; take, for example,

$$x \sim y \quad \text{if } x < y.$$

It is in the use of the adjective *equivalence* that we distinguish relationships which partition from those that do not - and the adjective is justified only when three conditions are met. For all elements of the set in question:

1. $x \sim x$: the reflexive property;
2. if $x \sim y$ then $y \sim x$: the symmetric property;
3. if $x \sim y$ and $y \sim z$ then $x \sim z$: the transitive property.

Immediately, we can see that $<$ fails with the first two conditions (and that $\leqslant$ just with the second) but that all three conditions are met with the first three examples.

It is a simple matter to show that any relation on a set which satisfies those three conditions must partition that set, and to do this we argue as follows.

Let the set be S. With the reflexive property, every element of S is contained in its own equivalence class, so all elements are accounted for. Now write the equivalence class containing x as X and suppose that $X \cap Y \neq \phi$. If $z \in X \cap Y$ then $z \sim x$ and $z \sim y$ and using the reflexive property, $x \sim z$ and $z \sim y$, and the transitive property $x \sim y$; the two equivalence classes are the same.

So, an equivalence relation partitions the set on which it is defined. Conversely, we may think of any partition of a set defining an unspecified equivalence relation with the equivalence classes precisely those partitions.

Now let us look at the detail of our use of equivalence relations.

On page 175 and 'equivalently' on page 223 we defined a relation on the set of real numbers by:

> Two real numbers α and β are said to be equivalent if there exist integers p, q, r, s with $|ps - qr| = 1$ so that $\beta = (p\alpha + q)/(r\alpha + s)$.

Let us now show that this satisfies the conditions of an equivalence relation.

1. For all α, $\alpha \sim \alpha$ since $\alpha = (1\alpha + 0)/(0\alpha + 1)$ and $|1 \times 1 - 0 \times 0| = 1$.

2. If $\alpha \sim \beta$, then $\beta = (p\alpha + q)/(r\alpha + s)$, where $|ps - qr| = 1$. This means that $\alpha = (s\beta - q)/(-r\beta + p)$ and we have that $|ps - (-q)(-r)| = 1$. Therefore, $\beta \sim \alpha$.

3. Finally, suppose that $\alpha \sim \beta$ and $\beta \sim \gamma$. Then, $\beta = (p\alpha + q)/(r\alpha + s)$ and $\gamma = (t\beta + u)/(v\beta + w)$, where $|ps - qr| = 1$ and $|tw - uv| = 1$.

   Therefore,

   $$\gamma = \frac{t(p\alpha + q)/(r\alpha + s) + u}{v(p\alpha + q)/(r\alpha + s) + w}$$
   $$= \frac{pt\alpha + qt + ru\alpha + su}{pv\alpha + qv + rw\alpha + sw} = \frac{(pt + ru)\alpha + (su + qt)}{(pv + rw)\alpha + (qv + sw)},$$

   where

   $$|(pt + ru)(qv + sw) - (su + qt)(pv + rw)|$$
   $$= |ps - qr||tw - uv| = 1.$$

It is clear that the rational numbers occupy one class, since if $\beta(\sim \alpha)$ is rational, so must be α.

On page 243 we defined the relation between two Cauchy sequences:

> Two Cauchy sequences $\{x_r\}$ and $\{x'_r\}$ are said to be equivalent (written $\{x_r\} \sim \{x'_r\}$) if, for every rational $\varepsilon > 0$, there exists a positive integer N so that $|x_r - x'_r| < \varepsilon$ for $r > N$; that is, eventually their difference approaches 0.

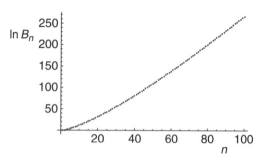

Figure E.1.

1. reflexivity is obvious

2. symmetry is obvious

3. Suppose that $\{x_r\} \sim \{x'_r\}$ and $\{x'_r\} \sim \{x''_r\}$, which means that $|x_r - x'_r| < \varepsilon$ and $|x'_r - x''_r| < \varepsilon$ for r sufficiently large; then $|x_r - x''_r| = |(x_r - x'_r) + (x'_r - x''_r)| \leqslant |x_r - x'_r| + |x'_r - x''_r| < 2\varepsilon$ for r sufficiently large, and we are done.

On page 248 we had defined the rational numbers as the set of equivalence classes of pairs of integers under the relation

$$(a, b) \sim (c, d) \quad \text{if and only if } ad = bc.$$

1. For all (a, b), $(a, b) \sim (a, b)$ since $ab = ab$.
2. If $(a, b) \sim (c, d)$ then it is clear that $(c, d) \sim (a, b)$.
3. If $(a, b) \sim (c, d)$ and $(c, d) \sim (e, f)$ then $ad = bc$ and $cf = de$. Thus, $a/b = c/d$ and $c/d = e/f$ and so $a/b = e/f$ and $af = be$ and we are done.

This formalizes the idea that, for example, $\frac{2}{3} = \frac{4}{6} = \frac{6}{9} = \dots$.

Finally, our interest has been with infinite sets but the concept of equivalence is meaningful if the set in question is finite and is particularly appealing if we adopt the use of partitions, suppressing the explicit use of equivalence relations. We now have a decent question to ask: How many partitions of a set of n elements are there?

If the set has just one element, $S = \{a\}$, then there is just one partition, the set itself.

If the set has two elements, $S = \{a, b\}$, then there are two partitions: $(\{a\} \{b\})$ and $\{a, b\}$.

Table E.1.

n	B_n
1	1
2	2
3	5
4	15
5	52
6	203
7	877
8	4,140
9	21,147
10	115,975
20	51,724,158,235,372

If the set has three elements, $S = \{a, b, c\}$ there are five partitions:

$$(\{a\}\{b\}\{c\}), (\{a\}\{b\ c\}), (\{b\}\{a\ c\}), (\{c\}\{a\ b\}), \{a\ b\ c\}.$$

The reader may be surprised at the nature of the general formula for the number of partitions (equivalence relations) possible for a set of n elements, B_n. It is given recursively by *Dobinski's formula* and defines the *Bell numbers*

$$B_{n+1} = \sum_{k=0}^{n} \binom{n}{k} B_k \quad \text{and} \quad B_0 = 1 : B_1 = 1$$

and explicitly in terms of an irrational number

$$B_n = \frac{1}{e} \sum_{k=0}^{\infty} \frac{k^n}{k!}.$$

Table E.1 given the first few values of the fast increasing B_n and figure E.1 provides a logarithmic plot of a few values more.

APPENDIX F

The Mean Value Theorem

One of the many significant ancient mathematicians whom we omitted in our chapter 2 romp through time and place was the Indian Vatasseri Parameshvara (1360–1425). One that we did mention was Bhaskara II, and it was in Parameshvara's commentary on Bhaskara's work that we find the essence of a result of calculus that was to find expression in 1691 in Rolle's Theorem. In modern notation, the result is:

> Let $f(x)$ be a real-valued function that is continuous on the closed interval $[a, b]$ and differentiable on the open interval (a, b), with $f(a) = f(b)$. Then there is some real number $c \in (a, b)$ such that $f'(c) = 0$.

A sketch provides overwhelming conviction that the result is true and it is a small matter of elementary real analysis to show that it is so. On page 184 we used its first generalization, the Mean Value Theorem, which in the form first given by Cauchy, states:

> Let $f(x)$ be a real-valued function that is continuous on the closed interval $[a, b]$ and differentiable on the open interval (a, b). Then there is some real number $c \in (a, b)$ such that $f(b) - f(a) = (b - a)f'(c)$.

The result is of wide-ranging importance yet its proof, given Rolle's Theorem, is trivial.

Define the function

$$g(x) = f(x) - \frac{f(b) - f(a)}{b - a}(x - a).$$

Then $g(x)$ has the same properties as $f(x)$ with the added feature that $g(a) = g(b) = 0$; it therefore satisfies the conditions for Rolle's Theorem. Since $g'(x) = f'(x) - (f(b) - f(a))/(b - a)$, there is a point $c \in (a, b)$ such that $g'(c) = 0$ and so $f'(c) = (f(b) - f(a))/(b - a)$ and we are done.

Index

Abel, Niels, 132, 235
Abu Kamil, 57
al-Karajī, 60
Al-Khwārizmī, 57, 253
algebraic irrationals, 180
Almagest, 65
analytic geometry, 77
Anderhub, J. H., 32
angle
    trisection of, 201
Apéry, Roger, 138
aperiodic tiling, 258
apotome, 50
Arago, Dominique François Jean,
    97
Archimedean spiral, 276
Archimedes, 49
Aristotle, 13
arithmetic axioms, 249
Arithmetica Integra, 74
Arithmetica infinitorum, 85
Arithmetical Theorem, 135
Aryabhata, 254
Axiom of Eudoxus, 40, 43, 49,
    167, 251

Beatty, Samuel, 261
Bell numbers, 293
Berlin Academy, 104
Bernoulli, Jacob, 96
Besicovitch's Constant, 233
Besicovitch, A. S., 233
Bézout's Identity, 166
Bhaskara II, 54
Binet formula, 221
binomial, 50
Blake, William, 152
Bolzano, Bernard, 237
Borel, Emil, 230
bride's chair, 45
Brouncker, Lord, 93
Burkert, Walter, 12

Caliph
    al-Ma'mūn, 56
    Harun al-Rashid, 56
    Omar ibn Khattab, 56
Cantor, Georg, 204, 239
Cantor–Heine–Méray
    Model, 242
Cartwright, Dame Mary, 113
Catalan constant, 112
catenery, 253
Cauchy
    distribution, 252
    sequence, 242
Cauchy, Augustin-Louis, 237
Cavalieri Principle, 86
Chaitin, Gregory, 231
Champernowne's
    Constant, 232
Champernowne, David, 232
Chaucer, 66
Cicero, 65
commensurability, 13
commensurable in square, 48
common elements, 134
complementary sequences, 262
complete ordered field, 250
completeness axiom, 250
constructible numbers, 203
continued fractions, 92, 183
Conway Constant, 136
Conway, John Horton, 133
Copeland, A. H., 233
Copeland–Erdős Constant, 233
coprime, 252
Cosmological Theorem, 135
countable set, 204
counter-earth, 14
Crelle, August, 236

Davis, Philip J., 277
De Divina Proportione, 74
De Morgan, Augustus, 47
de Stainville, M. J., 109

Dedekind
    cut, 241, 244
    Model, 244
Dedekind, Richard, 39, 239
Dehn, Max, 269
Delian Problem, 201
dense set, 167
Descartes, René, 77
Diehard test suite, 229
Diophantine approximation, 154
Dirichlet's Diophantine
    Approximation Theorem, 163,
    269
Dirichlet's Simultaneous
    Approximation Theorem, 270
Dirichlet, Johann Peter Gustav
    Lejeune, 162
Discourse on Method, 80
Dobinski's formula, 293
Drawer Principle, 162
Dudeney, Henry, 268
dunce's cap, 64
Duns Scotus, 64
duplication of the cube, 201
dynamical systems, 258
Dyson, Freeman, 184

Einstein's field equations, 253
equimultiples, 41
equivalence relation, 289, 290
Eratosthenes, 49
Erdős, Paul, 112, 233
Euclid, 38
Euclid's Elements, 10, 11, 21, 39
Eudemian Summary, 10, 14
Eudemus of Rhodes, 14
Eudoxus of Cnidus, 39, 41, 49
Eudoxus's method of exhaustion,
    49, 238
Euler's Constant, 263
Euler, Leonhard, 96, 182, 274
Euler-Mascheroni constant, 112
exotic elements, 134

factorial function, 274
Fermat's Last Theorem, 199
Fermat, Pierre de, 77
Fibonacci of Pisa, 60, 62
Fibonacci Sequence, 166
field, 250

Floor function, 113, 146, 157, 261,
    266
Flos, 63
Folium of Descartes, 84
Fourier, Joseph, 109
Fraenkel, Aviezri S., 267
Frederick II, 63
Frege, Gottlob, 251
Freiman's constant, 175

Galois, Évariste, 132
Gamma Function, 274
Gelfond Schneider Theorem, 198
Gelfond, Aleksandr, 184, 199
geometric
    curves, 80
    mean, 50
Golden Ratio, 28, 74, 174, 220,
    236, 255
Gray, Robert, 208
Great Year, 65
greatest common divisors, 166,
    213

Hall's Ray, 176
Hall, Marshall, 176
Hamilton, Sir William Rowan, 237
Hankel, Hermann, 238
Hardy, G. H., 32
harmonic
    oscillator, 253
    series, 138, 238
height of a polynomial, 205
Hermite Identity, 192
Hermite, Charles, 113, 190, 193
hexagram, 27
Hilbert, David, 11, 169, 198, 248
Hindu civilization, 53
Hippasus of Metapontum, 20
Hobbes, Thomas, 85
Horn Angle, 40
House of Wisdom, 56
Hurwitz, Adolf, 169

incommensurability, 48
indicator function, 257
irrational numbers, 29, 53, 63, 65,
    74, 77, 97, 165, 167, 172, 174,
    176, 199, 204, 209, 232, 233,
    237, 242, 245, 247
    addition of, 126

alternative definition of, 3
and rational approximation, 160
and the Littlewood Conjecture,
    172
and the quintic equation, 132
and the Zeta function, 153
as non-surd algebraic, 132
birth of, 9
Cantor's construction of, 209
Cauchy definition of, 237
denseness of, 167
elementary definition of, 2
generation of using
    polynomials, 131
nineteenth-century definition
    of, 239
non-existence of, 204
quadratic, 176
Stifel's view of them, 74
the most irrational, 219
Weierstrass view of, 239
Iwamoto, Y., 119

Johannes of Palermo, 63
Jowett, Benjamin, 29

Kant, Immanuel, 18
Khinchin, Aleksandr, 210
Klein, Felix, 169
Knorr, Wilbur R., 33, 47
Kronecker, Leopold, 204

La Disme, 76
Lady Isobel's Casket, 268
Lagrange Spectrum, 174
Lambek, Joachim (Jim), 265
Lambert, Johann Heinrich, 104
least common multiple, 145
Lehmer, Dick, 229
Leibniz, Gottfried Wilhelm, 182
Liber Abaci, 62, 74
Library of
    Alexandria, 11, 56
    Babel, 234
Life of Brian, The, 52
Lindemann, Ferdinand, 194
Lindemann–Weierstrass Theorem,
    200
Liouville Approximation Theorem,
    183
Liouville Number, 183, 186

Liouville, Joseph, 183
Littlewood, John Edensor, 172
logarithmic spiral, 253, 276
Look and Say Sequence, 133
Louis XIV, 286
Luther, Martin, 75

Madhava of Sangamagramma, 55
Markov
    numbers, 175, 223
    spectrum, 222
McCabe, Robert L., 33
mean, 20
    proportion, 50
Mean Value Theorem, 184, 294
mechanical curves, 80
medial, 50
Méray, Charles, 239
method of normals, 80
Minkowski, Hermann, 169
Moser, Leo, 265
Murray, James, 23

National Institute of Standards and
    Technology, 254
    Test Suite, 229
Newcomb, Simon, 252
Niven polynomial, 116
Niven, Ivan, 116
normal, 230
    distribution, 253
    in base b, 230
    numbers, 230
number of the beast, 75

Ohm, Martin, 236
Omar Khayyam, 61
    Rubaiyat, 61
Online Encyclopedia of Integer
    Sequences, 209
order axioms, 250
ordered field, 250
Oresme, Nicole, 66, 138, 259

Pacioli, Luca, 74
Patruno, Gregg N., 127
Penrose Tiles, 258
pentagram, 25
Père Lachaise, 153, 286
permeability of free space, 254
Perron, Oskar, 175

Pigeonhole Principle, 162, 226
Plato, 12, 21, 29
Plato's Dialogues, 29
Poincaré, Henri, 190, 239
Poinsot, Louis, 110
Poisson distribution, 253
Pope Leo X, 75
prime counting function, 147
Prime Number Theorem, 147
Proclus, 10, 15, 38
Profound Doctor, 66
Proteus, 90
Pythagoras
    constant, 1
    of Samos, 9, 10
Pythagoras's Theorem, 18, 38, 45

quadratic
    irrationals, 176, 186
    surd, 212
quadratrix, 81
quadrature, 85

rational
    line, 278
    point, 278
    roots theorem, 130
Rayleigh, Lord, 261
Riccati Equation, 102
Riccati, Count Jacopo Francesco,
    102
Riemann Hypothesis, 199
Rolle's Theorem, 294
Roth, Klaus, 181
Ruffini, Paolo, 132

Schneider, Theodor, 199
schnitt, 241
sexagesimal, 63
Siegel, Carl, 184, 199
Sierpiński, Waclaw, 231
similar figures, 36
simple
    continued fractions, 211
    pendulum, 252
simply normal in base b, 230
size of a polynomial, 208
Skolem, Thoralf, 268

Smith, Henry, 162
Spiral of Theodorus, 7, 35, 47, 272
squaring the circle, 15, 201
Śrīpati, 53
star polygon, 25
Steinhaus, Hugo, 259
Stevin, Simon, 76
Stifel, Michael, 74
Stupor Mundi, 63
Subtle Doctor, 64
Sum of Two Squares, 80
surd, 125
    etymology of, 57
Swineshead, Richard, 66

Tannery, Paul, 248
Thales of Miletus, 9, 10, 12, 16
Theodorus Constant, 277
Thomae, Carl, 251, 257
Thomas Bradwardine, 66
Thue, Axel, 184
transcendental
    functions, 81
    numbers, 183, 185, 197, 199,
        204, 210, 232
transuranic elements, 134
triangle inequality, 156

van der Poorten, Alfred, 139
van Pesch, J. G., 14
van Roomen, Adriaan, 70
Vesica Pisces, 28
Vièta, Francois, 70

Waclaw Sierpiński, 259
Wallis, John, 85, 93
Wantzel, Pierre, 202
Weierstrass, Karl, 200, 236
Weierstrass–Heine Model, 241
well-tempered musical scale, 254
Weyl, Hermann, 199
Wiles, Andrew, 199

Year of the Irrational, 239

zero
    etymology of, 62
Zeta function, 137